W9-CDX-865

A Journey Through Time

Exploring the Universe with the Hubble Space Telescope

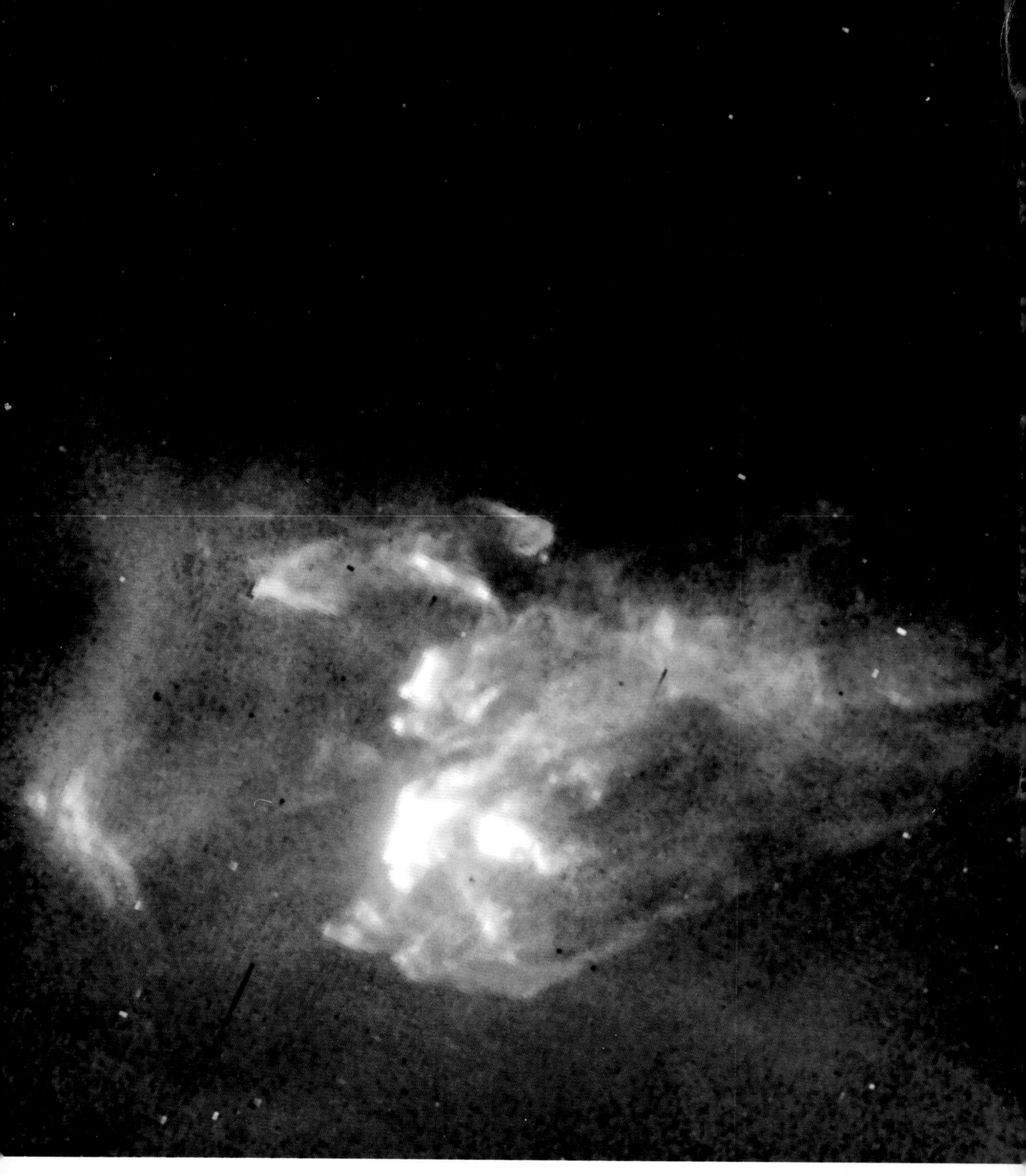

Above: A jet of gas ejected by a young star (HH1/HH2) lying 1,500 light-years away in the constellation Orion. *(NASA)*

A Journey Through Time

Exploring the Universe with the Hubble Space Telescope

BY

Jay Barbree and Martin Caidin

FOREWORD BY

Senator John H. Glenn, Jr.

DESIGNED BY William A. Sponn

PENGUIN STUDIO

Published by the Penguin Group
Penguin Books USA Inc., 375 Hudson Street,
New York, New York 10014, U.S.A.

Penguin Books Ltd, 27 Wrights Lane,
London W8 5TZ, England

Penguin Books Australia Ltd, Ringwood,
Victoria, Australia

Penguin Books Canada Ltd, 10 Alcorn Avenue,
Toronto, Ontario, Canada M4V 3B2

Penguin Books (N.Z.) Ltd, 182–190 Wairau Road,
Auckland 10, New Zealand

Penguin Books Ltd, Registered Offices:
Harmondsworth, Middlesex, England

First published in 1995 by Viking Penguin,
a division of Penguin Books USA Inc.

10 9 8 7 6 5 4 3 2

LIBRARY OF CONGRESS CATALOGING IN PUBLICATION DATA
Barbree, Jay.
A journey through time: exploring the universe with the Hubble Space Telescope / by Jay Barbree and Martin Caidin ; foreword by John H. Glenn, Jr.
p. cm.
Includes index.
ISBN 0-670-86018-2
1. Hubble Space Telescope. 2. Orbiting astronomical observatories.
3. Outer space—Exploration. I. Caidin, Martin. II. Title.
QB500.268.B37 1995
520—dc20 95-6758

Printed in the United States of America
Set in ITC Century Book
Designed by William A. Sponn

Now and then extraordinary people
step forward to do extraordinary things.
This book is for such people:
those who repaired the Hubble Space Telescope
370 miles above Earth.

Astronauts
Richard Covey,
Kenneth Bowersox,
Story Musgrave,
Claude Nicollier,
Kathryn Thornton,
Tom Akers,
and
the space-walking astronomer
Jeffrey Hoffman

SPECIAL ACKNOWLEDGMENT

All projects begin with a single idea. This book owes its birth to Peter Miller, who nurtured it to maturity.

ACKNOWLEDGMENTS

These pictures and words
were made possible by those
who stay focused on the heavens.

NASA
The Hubble Space Telescope Teams
Space Telescope Science Institute
U.S. Naval Observatory
Naval Research Laboratory
U.S. Geological Survey
and all of
Earth's observatories

With special thanks to
Hugh Harris, Ray Villard, Don Savage,
Mike Gentry, Jim Elliott, Diana Boles,
Bill Johnson, Linda Greenway, Nena Jones,
and
Margaret Persinger

CONTENTS

FOREWORD

by

Senator John H. Glenn, Jr.

First American Astronaut to Orbit Earth

When the Project Mercury team launched me into space in February 1962, I had little idea what I'd see up there. I remember being thrilled to be the first American to witness the first sunset in orbit. I told the ground, "This moment of twilight is simply beautiful. The sky is very black, with a thin glowing band of all the colors of the spectrum along the horizon." The sun went down at eighteen times its normal speed, followed by a slow but continuous reduction in light intensity. There were brilliant orange and blue layers that followed the curvature of the earth on either side of the sun that tapered gradually to a terminator point—of nothing.

Then it was the blackest of velvet nights. Stars went on forever for a distance I could barely fathom. I tried to add perspective to what I saw, tried to grasp the full size of our galaxy. The vastness of the universe at that moment was overwhelming. I was traveling at almost 5 miles per second (nearly 18,000 miles per hour), but what if I could travel at the speed of light—186,273 miles every second? Even at this almost incomprehensible speed, it would take me one hundred thousand years simply to travel from one end of the Milky Way to the other. Off in the distance I looked at other pinpoints of light and understood these were not individual stars but entire galaxies like our own, so far away they were barely visible.

No matter how much I wanted a powerful lens to bring them into full view, all I had was a small handheld camera, and in my spacesuit my gloved fingers kept fumbling with the film in weightless flight. If only I could capture the stunning vistas about me! It simply wasn't possible then.

But it is possible *now*, even to pursue billions of galaxies racing away from one another at millions of miles every hour because of the incredible rate of expansion of our universe.

We now have much more than a mere camera. We travel through nebulae, galaxies, clusters, exploding stars; we skim the edges of black holes, and we even collect pictures of an infant universe that existed only a short time after creation itself. We are doing it with a time machine we know as the Hubble Space Telescope.

Jay Barbree and Martin Caidin, two writers with seventy years' experience together in covering space exploration, are exceptionally qualified to write of the wonders revealed by Hubble. Barbree received an Emmy for NBC TV's coverage of the first landing of men on the moon. Caidin's motion picture about astronauts stranded in orbit, *Marooned*, received an Academy Award. They have filled these pages with Hubble's stunning photographs and momentous events. They show us how this telescope has altered forever how we view this magnificent universe that is our home.

From the beginning of time it is a marvelous journey, a breathless rush of creation and re-creation, moments of glory frozen in these remarkable scenes.

Join with them in this most exciting of all flights—*A Journey Through Time.*

INTRODUCTION

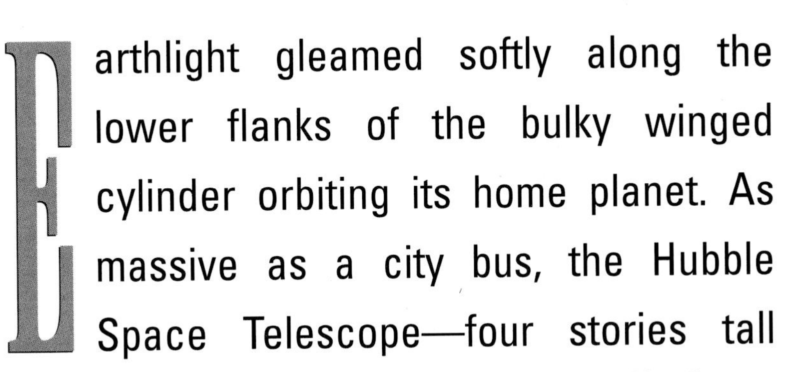

Earthlight gleamed softly along the lower flanks of the bulky winged cylinder orbiting its home planet. As massive as a city bus, the Hubble Space Telescope—four stories tall and weighing more than twelve tons—raced at 17,500 miles per hour around the dark side of Earth. The lights from below issued from brilliantly lit cities, forest fires, burning plumes of oil and natural gas wells; in the atmosphere itself, lightning bolts and streaks from meteors burned fiercely.

Despite being shrouded by the planet's shadow, Hubble reflected the bright pinpoints of stars gleaming everywhere beyond Earth's curving horizon. Jupiter, Mars, Saturn, and Venus hung in the velvet black sky like distant lighthouses.

Hubble was a dream started decades earlier. In 1946, Princeton astronomer Lyman Spitzer urged the U.S. government to build an orbiting space platform with revolutionary instruments to probe the universe, from nearby planets to the most distant stars. No matter how powerful the astronomical telescopes on Earth might be, they would never see clearly through our thick and pollution-muddied atmosphere. A telescope orbiting high above

Previous page: Shortly after its launch in April 1990, astronomers admitted the celebrated Hubble Space Telescope was seriously flawed; it is seen here in orbit over Madagascar. NASA came up with a plan to fix Hubble. *(NASA)*

Right: On the repair mission to the Hubble Space Telescope, the space shuttle Endeavour lifts off at 4:27 A.M., December 2, 1993. *(NASA)*

Earth could survey the heavens with unprecedented clarity.

But once in orbit, Hubble was sidetracked with flawed vision and shuddering vibration. Two months after the spectacular fiery ascent from Cape Canaveral in April 1990, embarrassed astronomers admitted that the telescope's goals were seriously compromised. Some systems worked well, but not the most critical scientific packages.

The most celebrated telescope since Galileo assembled his first optical instrument was sending blurred images back to Earth. Astronomers admitted that the most modern telescope was essentially a flop. Critics castigated NASA and the team of Hubble astronomers for spending the greatest amount of money on any one astronomical project, yet returning for the investment results assailed as "pitiful."

NASA—and the astronomical community—had been vainglorious in its promises that Hubble would produce an unprecedented stream of sharp pictures of light traveling from the beginning, or very nearly the beginning, of time. The promise collapsed like a punctured balloon.

Hubble's primary mirror worked dismally; the telescope's electronics sent back to Earth pictures that were fuzzier than snapshots taken by the shaking hands of a child with Kodak's cheapest box Brownie. That precious eight-foot primary mirror, which had required five years of grinding and polishing by the best teams in the country, was flawed.

Its edges had been ground two-millionths of a meter too flat.

John Troeltzsch of Ball Corporation, which built corrective optics to eliminate the blurred pictures of the primary mirror, explains, "The mirror is ten-thousandths of an inch too flat."

That sounds insignificant. It is only one-fiftieth of the diameter of a human hair, which means it is invisible to the human eye.

But in the optical world of mirrors and lenses built to see twelve billion miles across the universe, that amount of error was an enormous distance. So Hubble became instant grist for late-night television comedians and a butt of ridicule for American science.

Quick to defend the initial positive but blurred results of Hubble, NASA's Dr. Edward J. Weiler insisted that the early pictures from Hubble were "outstanding results in themselves." With rising optimism, he added, "They are nonetheless just the appetizer for the main course to come."

Much more to the point was the comment from J. Jeff Hester of Arizona State University's astronomy program, where scientists were hard at work to come up with a fix for the myopic observa-

Right: Astronaut Story Musgrave glides gently and carefully on the penultimate space walk of the mission to perform precision surgical repair, replacement, and modification to the twelve-ton space observatory, with Hubble anchored securely to its berthing platform in the shuttle's cargo bay. The huge telescope gently orbits the planet at 17,500 miles an hour. *(NASA)*

tory. "Looking at something with the Hubble right now," he said dryly, "is like looking through a piece of frosted glass."

That cut right through the public relations hoopla that was doing its best to minimize failure while promising bonanzas for the future. Chagrined project managers planned risky schemes to eliminate the blunders. COSTAR (Corrective Optics Space Telescope Axial Replacement), a "fixit effort" to be done with mirrors, was born. No dry humor here.

COSTAR was a rectangular box the size of a telephone booth. It weighed 650 pounds, and it contained ten mortised mirrors. It was about as close to technical magic as one could get. Each of its ten mirrors was no larger than a man's thumbnail!

Spacewalking astronauts would fly a shuttle to orbital rendezvous with Hubble, grasp the huge observatory tightly with the shuttle's 50-foot robotic arm, and then begin their "save the Hubble" week in space. Among other repairs, they would slide COSTAR in place within the main structure, where the mirrors would shorten the beam of light images captured by the flawed edges of the primary mirror. By shortening the beam of light exactly two-millionths of a meter, Hubble would be able to focus accurately and correctly. Once the ten mirrors were in place, astronomers in ground control would transmit aiming and focusing instructions to the orbiting observatory. They would (they hoped) focus and adjust COSTAR's mirrors by tipping them into thousands of different positions.

But more than COSTAR was necessary to bring Hubble back to pristine performance. New forty-foot solar panels would replace those that shuddered sixteen times a day when the ship passed between day and night, through temperature changes of 500°F. Two magnetometers that had lost their precision attitude control would be removed to make way for new ones. A series of critical gyroscopes that point Hubble on command had either failed or were failing; new gyros would be installed. COSTAR would be eased by the spacewalkers into its housing. A new computer would be added to Hubble to eliminate "electronic memory lapses" and increase the space telescope's reliability. And finally, the spacewalking astronauts would repair flawed relays in the spectrograph that scanned the radiations of the universe.

Space shuttle Endeavour departed Earth on Mission 61 at 4:27 A.M. eastern time on December 2, 1993. Seven astronauts along with 15,000 pounds of tools, special equipment, and supplies were borne aloft on twin pillars of blazing rocket fire that shattered the quiet of east Florida's sleeping communities with earsplitting thunder.

When Endeavour reached Hubble's orbit, the astronauts found and captured the observatory with the shuttle's robotic arm. Then the spacewalkers, equipped with pressure suits and working in pairs, went through an astonishing week of giving the crippled telescope new life and sparkling accuracy.

Floating a constant 375 miles above a curving horizon, looking like living snowmen, the spacewalkers performed weightless ballets to make their repairs. It was a feat unparalleled in history, surgeons of the new age operating beneath a star-filled theater. There had not been so much rapt attention from billions of people on Earth since the first man walked on the moon. Awed television viewers on Earth were able to look over the shoulders of the astronauts as live TV cameras followed almost their every move.

Performing microsurgery on Hubble's systems, as well as moving bulky and cumbersome equipment into the right slot at the right speed and with perfect aim, was like trying to weave a frond basket wearing thick mittens. Working in pairs, the spacewalkers performed one intricate maneuver after another, until finally they installed COSTAR with its ten small mirrors that would give Hubble 20-20 vision.

The astronauts carried out their eleven major repairs while earthbound scientists sweated out every move. One bad move could wreck the mission, damage Hubble beyond repair. Yet, from changing fuses to sliding the refrigerator-sized COSTAR into the telescope's bowels with less than an inch of room to spare, they performed with astonishing, nonstop perfection. It all went so smoothly, even against terrible odds, that COSTAR's lead engineer, Jim Crocker, said with a huge grin, "We felt like the Maytag repairman."

Finally the nerve-racking mission was nearly complete. The old solar panels were removed from their securing masts. Spacewalker Kathryn Thornton, feet secured to the end of a long robotic arm, held the twisted panels in her hands and pushed them away in a tantalizing slow-motion ejection. The pilots aimed the shuttle's rocket motors at the old solar wings. Streaming gases struck the golden shapes; they flapped eerily up and down, looking like mankind's first space bird. Their orbit now slowed, they would soon fall back into Earth's atmosphere, ending their life in a streak of short-lived flame.

Eight days passed. Endeavour and her crew were back on Earth. Hubble managers waited fretfully to determine if the space surgery was as successful as it promised.

Then the last fetters of Hubble were removed. Power was fed

Left: Floating on the end of the shuttle's robotic arm at the top of the four-story Hubble Space Telescope, astronauts Story Musgrave (top) and Jeffrey Hoffman are seen above the west coast of Australia. *(NASA)*

Above: The Hubble Space Telescope, released from Endeavour, following a week and a half berthed in the shuttle's cargo bay. Part of Earth's horizon is visible in the lower right. *(NASA)*

Below: A fish-eye lens captures the Hubble Space Telescope and Earth below (showing the Australian landmass) during astronauts Story Musgrave and Jeffrey Hoffman's final repair space walk. Musgrave can be seen at the bottom of the huge telescope. *(NASA)*

to controls and instruments. The space observatory moved through its checkout commands with its thumb up. Hubble was again alive. Scientists gathered before a television monitor in the Space Telescope Science Institute in Baltimore, Maryland. The control center had all the tension of a maternity ward waiting room.

Light surged through the screen, flickered, then steadied. There—the first image from high above the world. *Star AGK +81 D226* . . . clear, sharp, *beautiful*. For moments silence gripped the room, then tension evaporated in tumultuous back-slapping, applause, cheering. Astronomers hugged one another fiercely, joyfully.

One astronomer spoke for them all. "It was like giving life again to something stillborn—and now it's starting to sing and dance."

Endeavour's spacewalkers had corrected Hubble's vision to even greater sharpness and clarity than its creators had ever hoped: a resolution more than ten times sharper than the most powerful telescopes on Earth. With its new eyes and equipment, it could scan with extraordinary clarity an area of space a thousand times greater than that covered by the biggest telescopes ever built.

"Beyond our wildest expectations," said beaming Hubble scientist Ed Weiler. NASA astronomers explained to awed newspeople that if they set up the main camera from Hubble in Washington, D.C., it could photograph clearly, and separate into two images, two fireflies only nine feet apart—in Japan!

From nearsightedness to super vision, this was the new Hubble. An NBC correspondent laughed and said to a companion, "It's amazing what you can do with a $629 million pair of contact lenses."

Before the Endeavour rescue-and-repair flight, Hubble could see out to four billion light-years away from Earth. Now Hubble's "vision reach" had tripled to 12 billion miles, and it promised to increase its clarity as engineers and astronomers tweaked the space observatory's systems.

Now there were questions to be answered about Hubble's multibillion-dollar probes back through time, when the universe was smaller, expanding rapidly, and ever-changing. We went to "the source," no less than Story Musgrave, payload commander and lead spacewalker on Endeavour's repair flight. Our first question: "What do you expect Hubble to tell us?"

"Hubble is delivering, right now, even more than was ever expected," Musgrave told us. "The new planetary cameras are state-of-the-art and they have their own corrections built in. They work far better than we ever hoped.

"How far back will we see in time? Well, time is one thing and the scientists are working on what is called the Hubble Constant. This is a critical cosmological parameter. It's a number that relates the velocity of an object in space—it's velocity and it's distance—that may better define the age of the universe. That's a real bone of contention right now. Is the universe ten billion years old? Twenty billion? We need better definition."

Then we talked with Story Musgrave's spacewalking partner, Jeffrey Hoffman, the only astronomer who climbed all over the four-story telescope to repair it in orbit, and we asked him, "What do *you* expect to learn from Hubble?"

The astronomer-astronaut went right to the quick. "The whole dream," Hoffman said slowly, "was to push back the frontiers of how far we can see out really towards the edge of the universe. That means backwards towards the beginning of time."

"Do you really expect to see back to the beginning of time?" we asked.

"Well, the beginning of time presents its own problems. You can't get back earlier than a certain phase, looking with a telescope, because there was a time when the universe was so dense that light doesn't escape that density. That's this cosmic background, the microwave radiation that pervades the whole universe. There's a limit even to the Hubble telescope, but it's going to take us back much farther than we have ever been able to see before.

"Hubble is going to open up a totally new picture on the early history of the universe. I'm looking forward to looking deep inside the galaxies, actually into a black hole, and most of all, I want to see what we can do in terms of locating planetary systems being formed. This is something human beings have speculated on for as long as we have thought about life, origins of life." Hoffman smiled. "Are we alone?"

"What's your best hunch?" we asked.

"I suspect there are Earths all over the place. Cosmic chemistry seems to work the same no matter where we look. I think there's life in other places. We need," he said flatly, "to keep going out and find out what's out there."

The possibilities offered by Hubble fire the imagination, because the answers to many questions may now be within our grasp.

M100 Galactic Nucleus

Hubble Space Telescope
Wide Field Planetary Camera 2

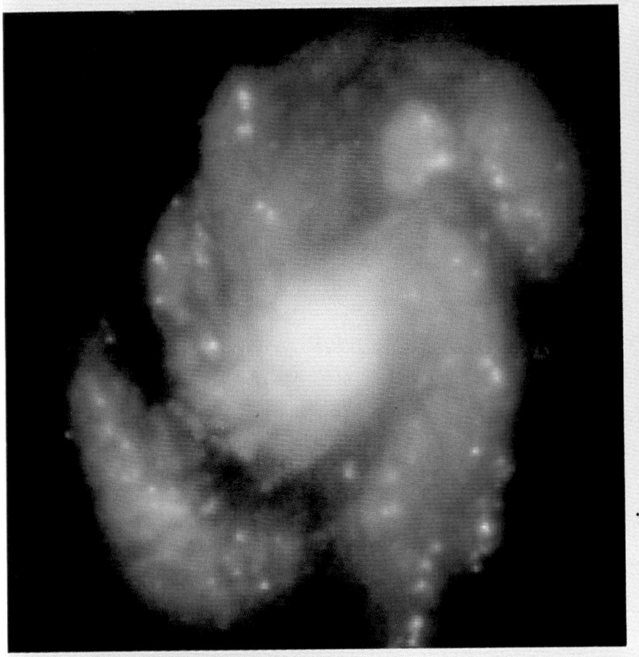

Wide Field Planetary Camera 1

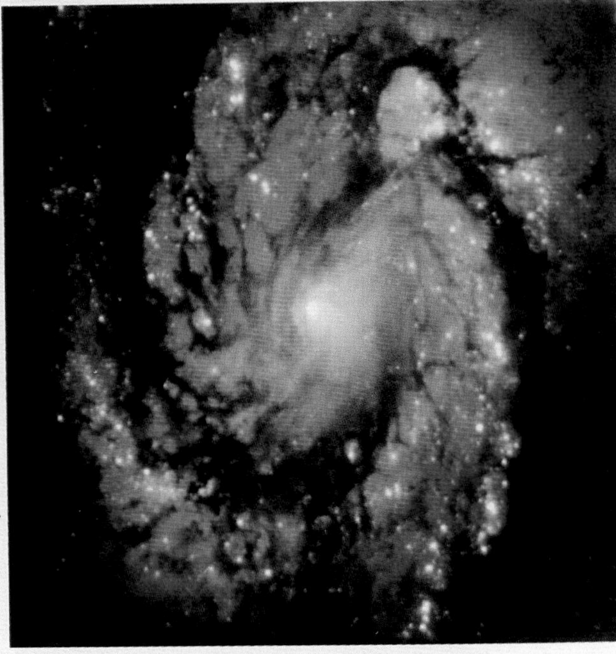

Wide Field Planetary Camera 2

Above: Joyful astronomers hugged one another as they saw this "before and after" by the Hubble Space Telescope. At left is the best picture that Hubble could provide when it was first released into Earth orbit. Spherical aberration problems with faulty optics produced the best pictures ever taken of the star Melnick 34 within the Large Magellanic Cloud—but far from what astronomers hoped to achieve. Then, immediately after the brilliant "fixit mission," came the picture on the right. *(NASA)*

Will we find planets with life on them? With intelligent life-forms? How do we communicate with them, and them with us?

Why are there single stars that burn with such intensity they're brighter than all the starlight, combined, from a billion suns?

How far back will we go in time before we reach that universe that is so much smaller, and denser, that light can't even move through . . . whatever it is?

What *is* gravity?

Is the speed of light really the ultimate velocity? Or will we find unanticipated matter and energy that travel many times the speed of light?

What happens to the trillions of tons of matter that vanish into the maw of the black hole?

What are the white gushers in space pouring vast amounts of subatomic particles into our universe—with no identifiable source or known reason?

If there truly was a beginning to our universe, the Big Bang, will we ever know how dense, how large it was, or even what created the Cosmic Egg that science postulates exploded to create our universe?

Right: One of the first stars photographed by Hubble after its repairs was the mighty stellar maverick of the universe Eta Carinea. Ten thousand years ago, five thousand years before the Pyramids were built, when the first farmers of our world were just learning to sow crops and store food through long winters, Eta Carinea "blew its stellar stack."

The image of the exploding star, traveling at 186,000 miles per second, took 10,000 years to reach us. Eta Carinea flashed so wildly in the heavens, it became the second brightest star studied by astronomers. *(NASA)*

If space is infinite, how can we reach an end to infinity?

If our universe is expanding, as certainly it seems to be according to every study and test astronomers make, will that expansion go on forever—or will the universe one day end its outward reach and begin to contract, finally ending jammed into another Cosmic Egg?

If we do learn how to travel faster than light, and launch journeys to other planets that have intelligent life, much older than our civilization, how will we react to being considered as little more than primitives?

The Egyptians built the pyramids and great temples. The ancient Britons built Stonehenge and amassed astonishingly accurate astronomical data. Hindus and Aztecs created the concept of zero and the decimal point, and made possible the world of mathematics that thrust us into the future. Americans led the way in aerial flight, and then broke the chains of gravity to walk and ride on the moon.

Now, through Hubble, our scientists begin the next great march into the future. We might yet come to understand our place in the order of being.

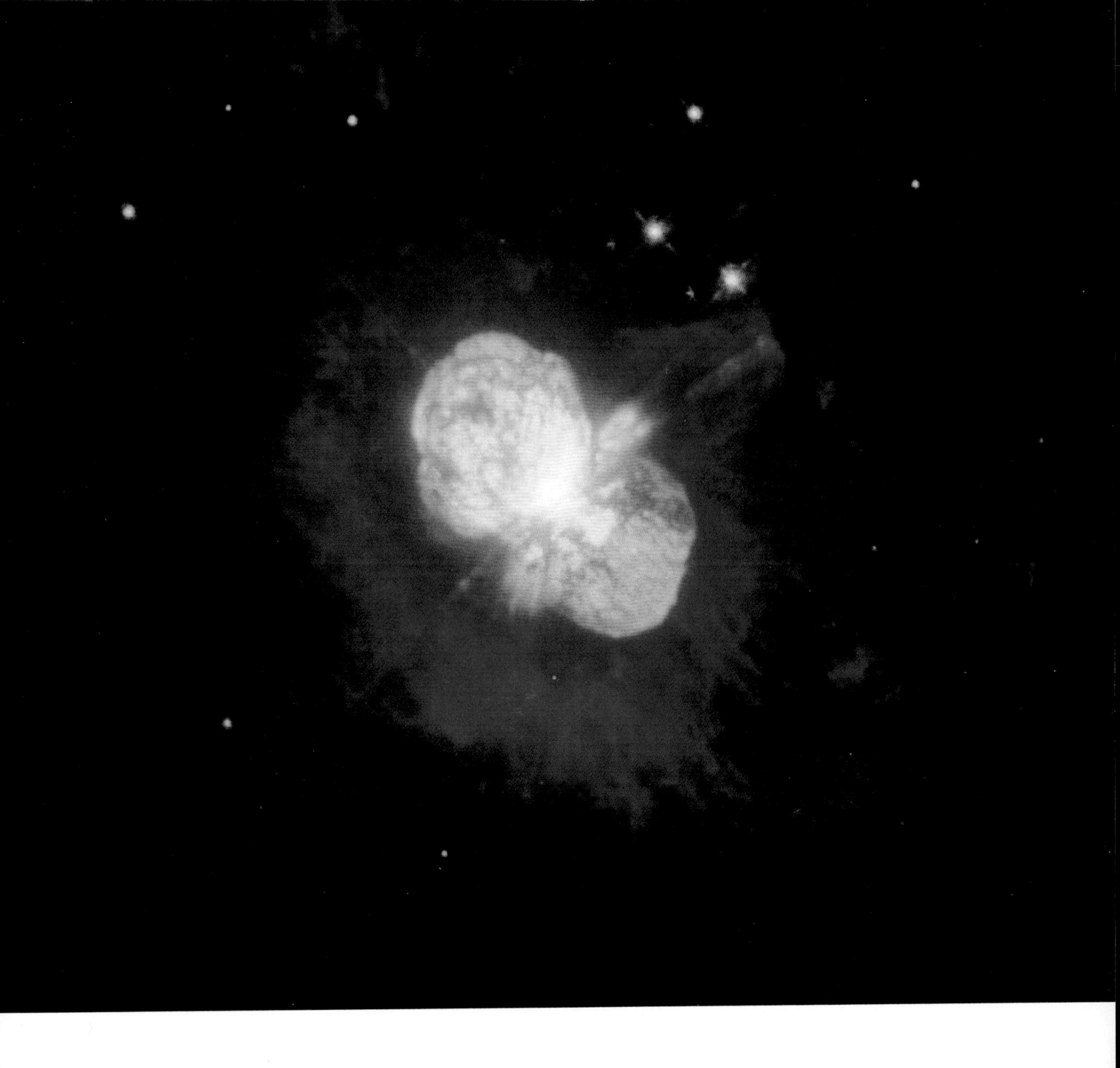

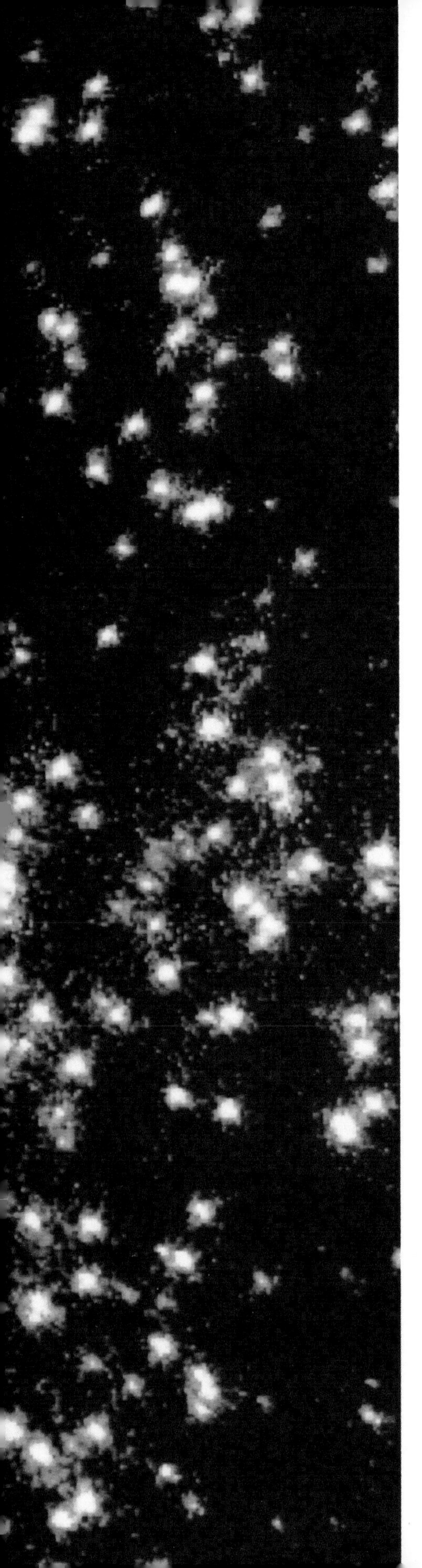

1 FIRST LIGHT

Before our universe existed, there was no past, present, or future.

Space did not exist, nor did vacuum. There was nothing—even dark cannot exist without light.

Scientists refer to the moment of creation as the Big Bang. They have nothing else to go by. Before the universe existed, they theorize, everything that ever was or will be, including time, was condensed into a single mote, which had to explode in order to create the universe. For want of a better term, they call this ultimate singularity the Cosmic Egg.

Since it has never been seen, nor ever will be seen, the Cosmic Egg cannot have any identifiable shape or size. It is a matter of faith deduced by looking as far as possible back in time to what scientists believe was a cataclysmic blast. It was the ultimate explosion, but it was never seen or heard. We know only its aftereffects.

By looking back through time with the Hubble Space Telescope to study the infant universe, astronomers may finally prove everything began when this mote shattered.

If the theory is true, the Big Bang is simply Time Zero—the first instant of existence, when the universe in which we live was born.

Previous page: The core of the globular cluster 47 Tucanae is seen here in a photograph by Hubble.

Astronomers say blue stragglers may evolve from "old age" back to a hotter and brighter youth in the infant universe through stellar collisions and mergers. This high concentration of blue stragglers toward the core of 47 Tucanae suggests they are significantly more massive than most of the cluster's stars. Some of the blue stragglers could really be double star systems. *(NASA)*

Right: This NASA artist's concept depicts crucial periods in the development of the universe according to inflationary cosmology, from a tiny fraction of a second after the Big Bang up to the universe as we see it today (15 billion years later). *(NASA)*

Space itself—within which everything else will happen—was also created. Energy burst forth as heat, light, pressure, and electromagnetic and nuclear forces erupting in all directions from the beginning of time.

As best as can be determined by our finest instruments and computers, the instant in which this took place is too small to be measured. As an example, in the explosion of an atomic bomb it takes plutonium a millionth of a second to split and release its pent-up nuclear fire. That millionth of a second, compared to the instant of the Big Bang, is perhaps equal to a century. But that is the best science can give us.

Then, still instantly—by the means with which we measure the passage of time—there is now mass in the form of subatomic particles, searing heat so great the universe is opaque, like white blindness. There is now gravity. No one has any idea of how gravity came to be, nor do we know today what gravity is. We measure its effect, we rely upon gravity as the engine that drives the universe, we know what it does, but we have not the faintest idea of what it is or how or why it works the way it does. But it is there in a violently exploding universe.

In a billionth of a second as we now measure time, or perhaps the first three minutes of existence, the physical laws that govern cause and effect started to permeate the newborn universe. Scientists calculate that in this indefinable tick of time, subatomic particles and electromagnetic radiation flashed from the point of explosion for trillions of miles.

That in itself is a contradiction of the physical laws in which we live today. Einsteinian equations state that nothing can travel faster than the speed of light. Anything rushing through space, newly created then or today, increases its mass as it increases velocity. It must somehow be impelled to accelerate. But when it reaches the speed of light, whether it be an atom or a spaceship, the mass of the object becomes infinite, beyond all measurement. Being infinite, it needs more than infinite energy to accelerate to greater speed. Ergo, the speed of light is the absolute limit of movement through space.

Light itself *might* have infinite mass. Since light—photons—changes back and forth from quanta (mass) to wave (pure energy), NOBODY KNOWS IF IT HAS EITHER INFINITE MASS OR NO MASS AT ALL AT THE ONLY SPEED AT WHICH IT TRAVELS: 186,273 miles per second.

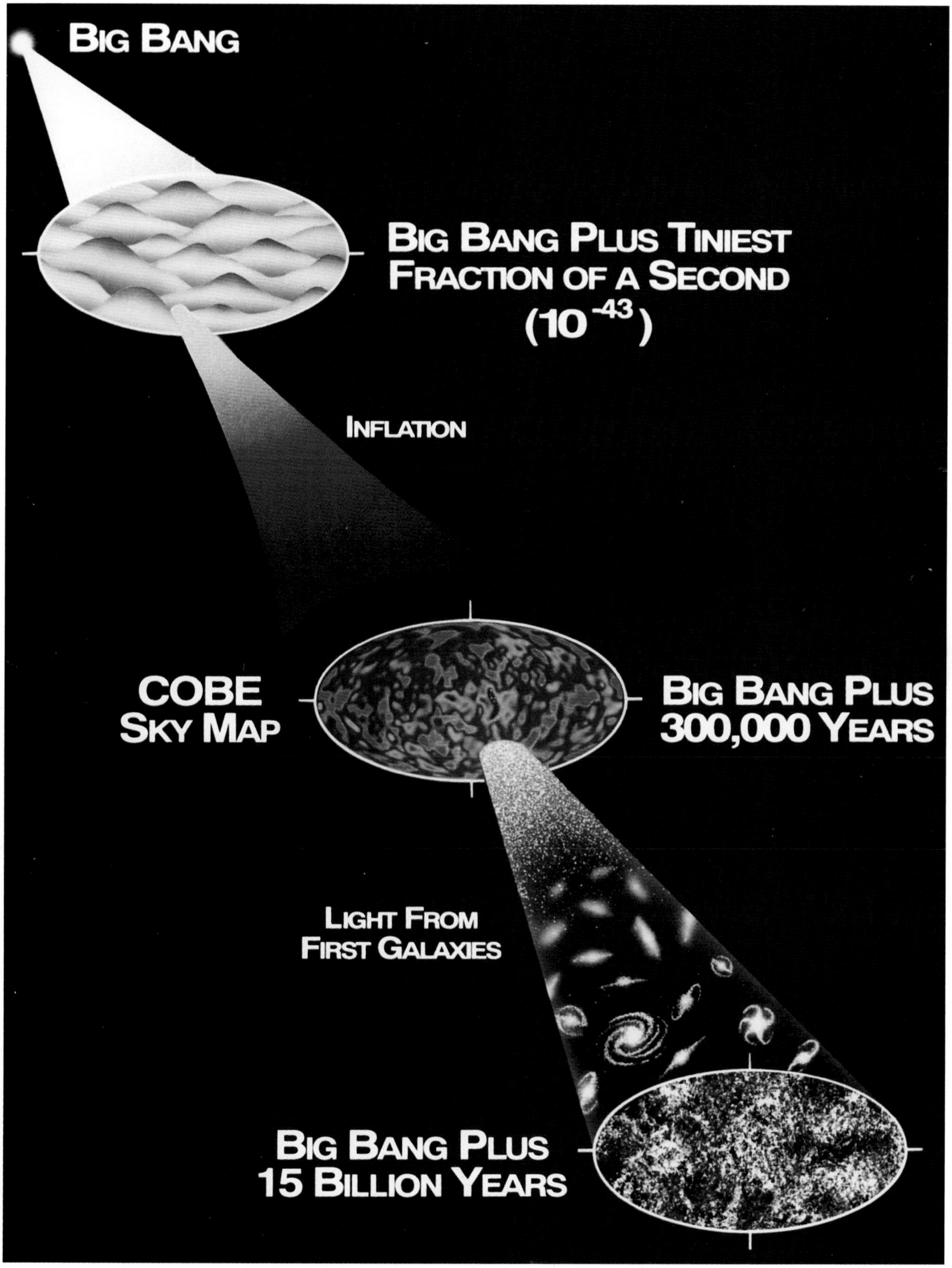

BIG BANG

BIG BANG PLUS TINIEST FRACTION OF A SECOND (10^{-43})

INFLATION

COBE SKY MAP

BIG BANG PLUS 300,000 YEARS

LIGHT FROM FIRST GALAXIES

BIG BANG PLUS 15 BILLION YEARS

But clearly not in the first moments of creation. Some scientists insist that at the first instant after Time Zero, the entire universe came into being instantly.

If what we know of our universe today is true, then in the earliest moments of time, every element that would one day exist—heat, light, mass, hydrogen, helium, the vast and swirling family of atomic and subnuclear particles—hurtled through newly created space at many times faster than the speed of light. We do not know when the limiting laws of velocity came into being.

Gravity appeared as mysteriously as everything else, forcing the swirling clouds of subnuclear gas to form huge clumps of elementary matter in space. These clumps became atomic elements. Gravity was the sticky glue to keep them bound in ever-larger shapes and mass. The engine to drive the universe worked at full speed. Subatomic particles merged, smashed together. Atoms were rearranged by collision and gripped tightly by gravity, and heavier new elements formed. These were the building blocks of *our* universe to come.

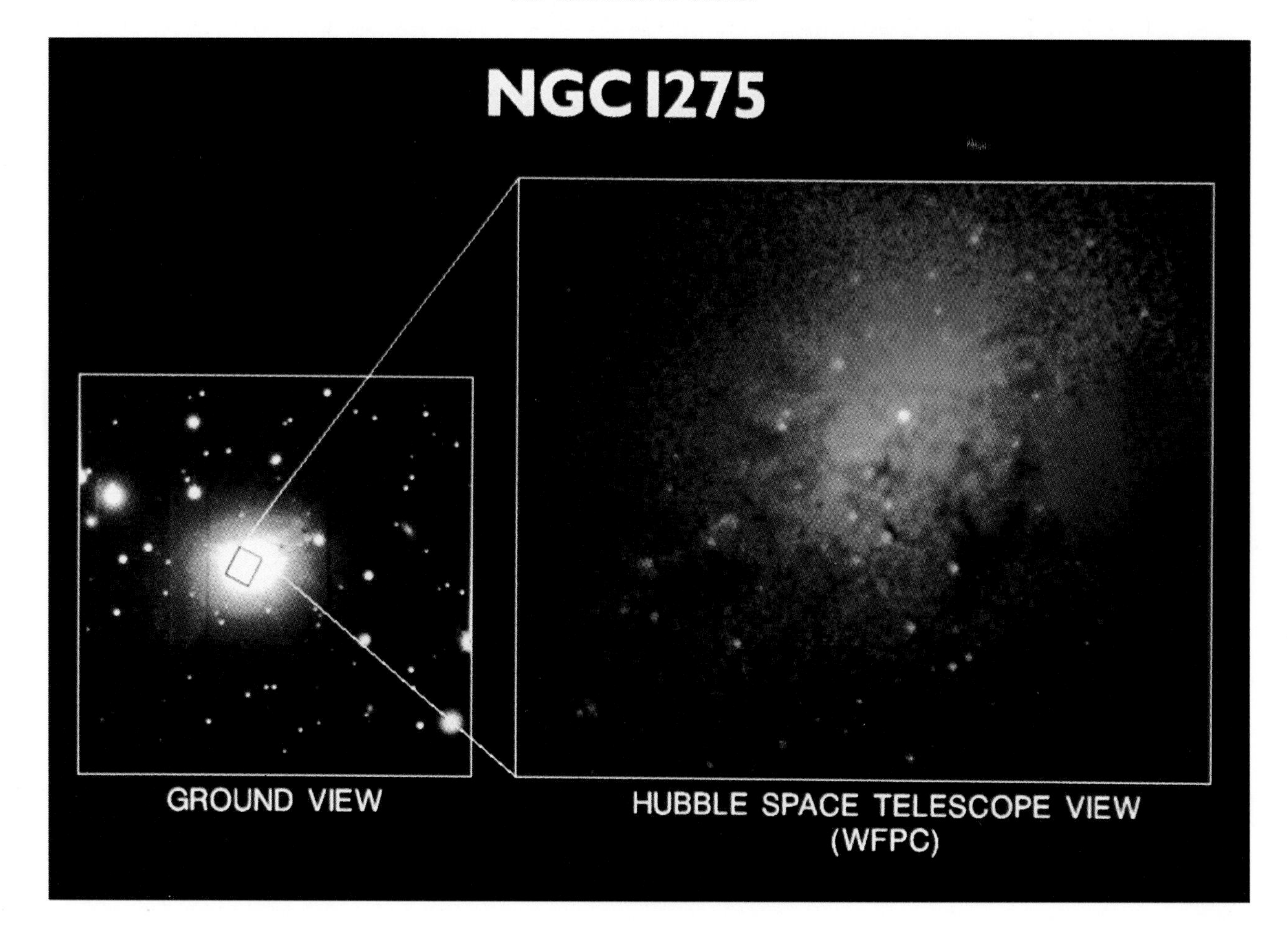

Below: On the left is a ground-based image of the giant elliptical galaxy NGC 1275, taken by the Kitt Peak National Observatory.

On the right, the Hubble's high resolution reveals individual clusters of stars that appear as bright blue dots. These globular clusters contain young stars much like those in the early formation of the universe. *(NASA)*

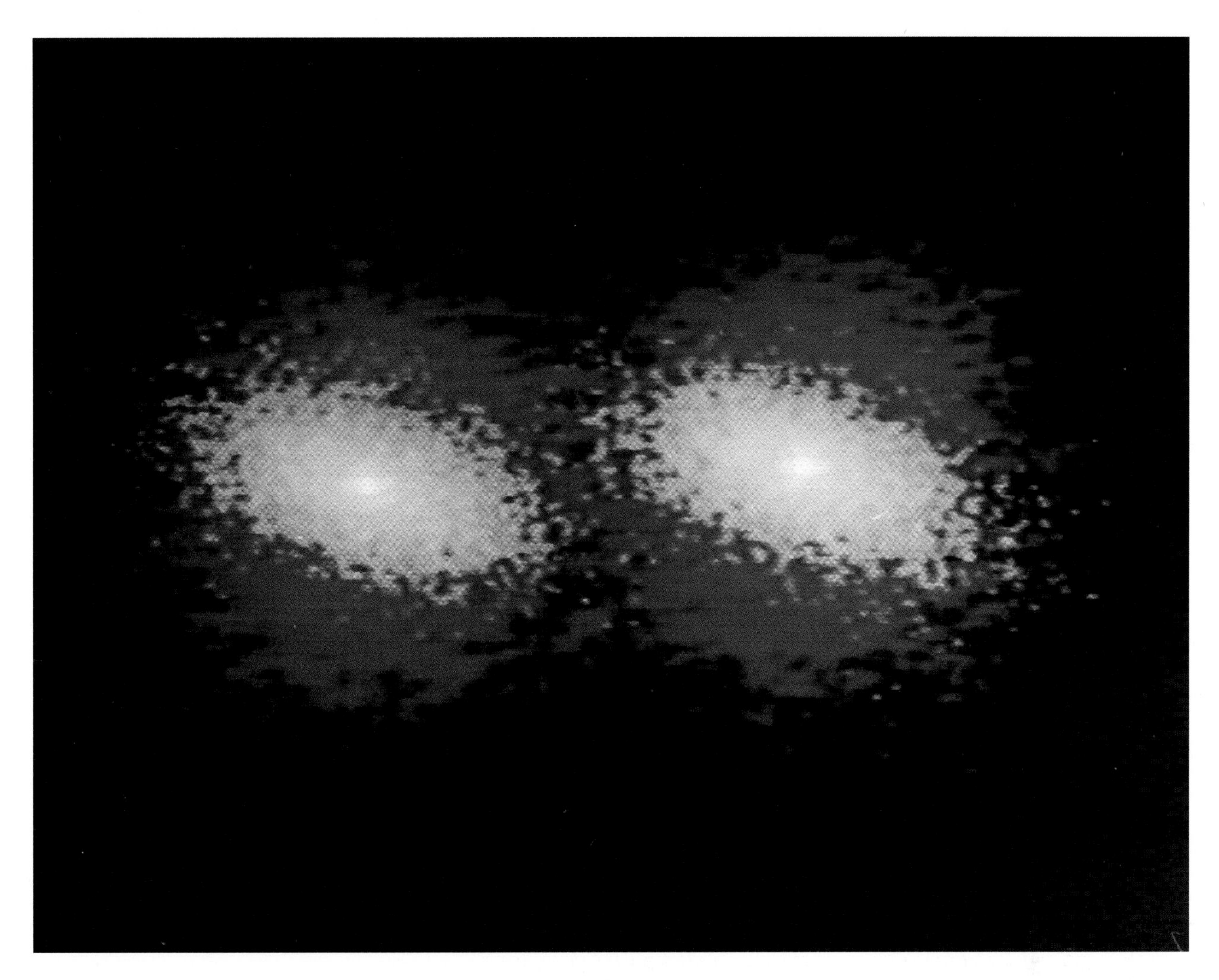

Above: In this photograph, the computer has taken information sent back by Hubble and simulated two spiral galaxies beginning to merge or collide. Such encounters may be responsible for creating elliptical galaxies. Astronomers suspect such collisions may trigger the formation of globular star clusters. *(NASA)*

At some time in this infant universe—at a moment scientists have yet to pinpoint—the outrushing debris of the Big Bang cooled. As matter collected in ever-larger clumps, it attracted through gravitational pull the swirling gases and particles through which it moved. Each clump was like a powerful magnet moving through an ocean of building material pouring in from all sides.

Gases and atoms jammed together. Atoms formed molecules. The temperature permeating the universe cooled. Out of this gruel the thin clumps joined in ever-increasing mass. Clumps became spherical shapes, their edges clear and defined.

Infalling matter has a natural tendency to turn, to spin slowly and then accelerate as it nears a common center. The raw material from which stars would be born flowed inward.

This is the time of the protostars. The amassing spheres that soon became so vast they heated up from contraction and the pressure of gravity driven inward.

A new law exerted itself—the conservation of angular momentum. Matter that swirls inward condenses as it spins. It is like

Right: Hubble found the Orion Nebula to be a "stellar nursery"—a region where new stars are forming out of interstellar gas. As shown here, the emission from the nebula is powered by the intense ultraviolet light from a cluster of particularly hot and luminous stars. The sulfur emission seen in this photograph is coming from the region where the light from these stars is "boiling off" material from the face of the dense cloud. This is the very cloud from which the hot stars formed, and is known to harbor additional ongoing star formation. *(NASA)*

an ice skater in a whirling pirouette; as she brings in her arms, holding them tightly against her body, she spins in a blur. Skater or aborning star, the law works the same.

Deep within the protostars the fury of nuclear fire began. Trillions of tons of hydrogen shifted atomic structure to fuse into helium. Each tiny transformation released a ghostly spasm of energy. The rush of hydrogen into helium accelerated with blinding speed. Hellish flames burst outward as protostars became blazing suns. Their tremendous internal energy threatened to blow them apart. The outward blast of stars burning furiously, radiating immense waves of energy, turned on the lights of the universe.

Everything would have blown apart were it not for gravity. As furiously as the new stars pushed outward, gravity rammed them inward. The explosive stellar pressure held in superb balance. Many stars consumed energy with fearsome speed. In several thousand years they burned to dimmed hulks. Those that were held in balance between thermonuclear fire and gravity would burn for billions of years.

Different stars, depending on their size and mass and the chemical elements being stoked in their deepest fires, burned in a fantasy of colors. Blue, white, yellow, red, orange—the colors everywhere were a stellar kaleidoscope.

Some stars drew together by mutual gravitational attraction to form dazzling binaries, star pairs revolving one about the other. Other new suns gathered in groups of three of four. Still more smashed together.

Still other stars, their fire-and-gravity balance unchecked, eventually exploded. Those titanic blasts hurled forth a vast potpourri of heavy elements that formed other stars.

Billions of stars, then trillions, moved in a stately celestial pirouette toward a common center where gravity dominated their every move. They began to wheel and spiral about that center to create the cities of stars—the galaxies.

Still creation marched on. Just as stars gathered in clusters and galaxies, so even the galaxies began to assemble as galactic families, so huge it took light, moving at 186,273 miles a second, more than a billion years to cross from one side to the other of the galactic structure.

The flotsam of material ejected by the collision and explosion of millions of stars flooded space with gases, dust, light, heavy atomic nuclei, and radiated energy. This stellar debris drew together in ever-larger masses. Gravity would finally squeeze them into spherical shapes.

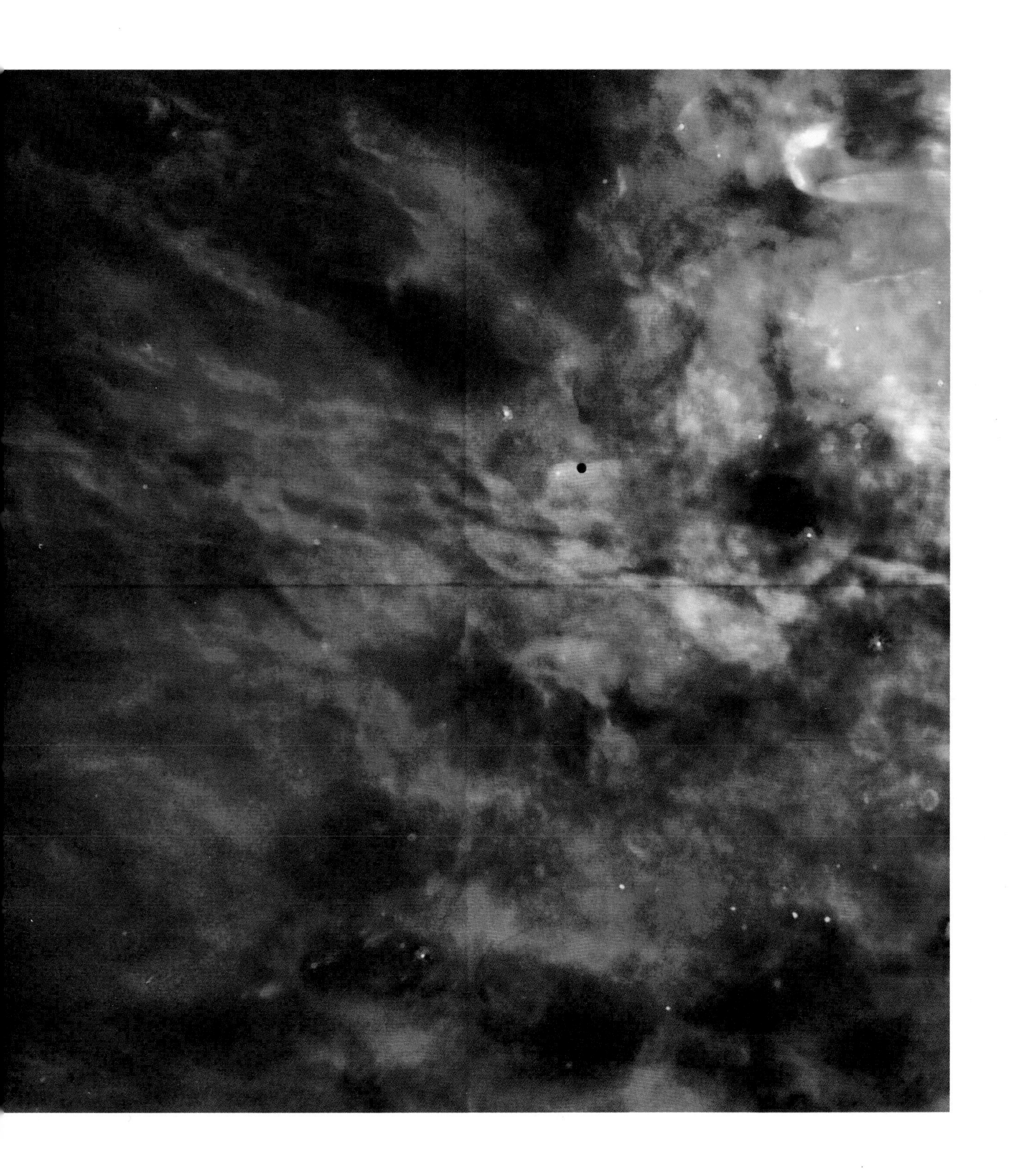

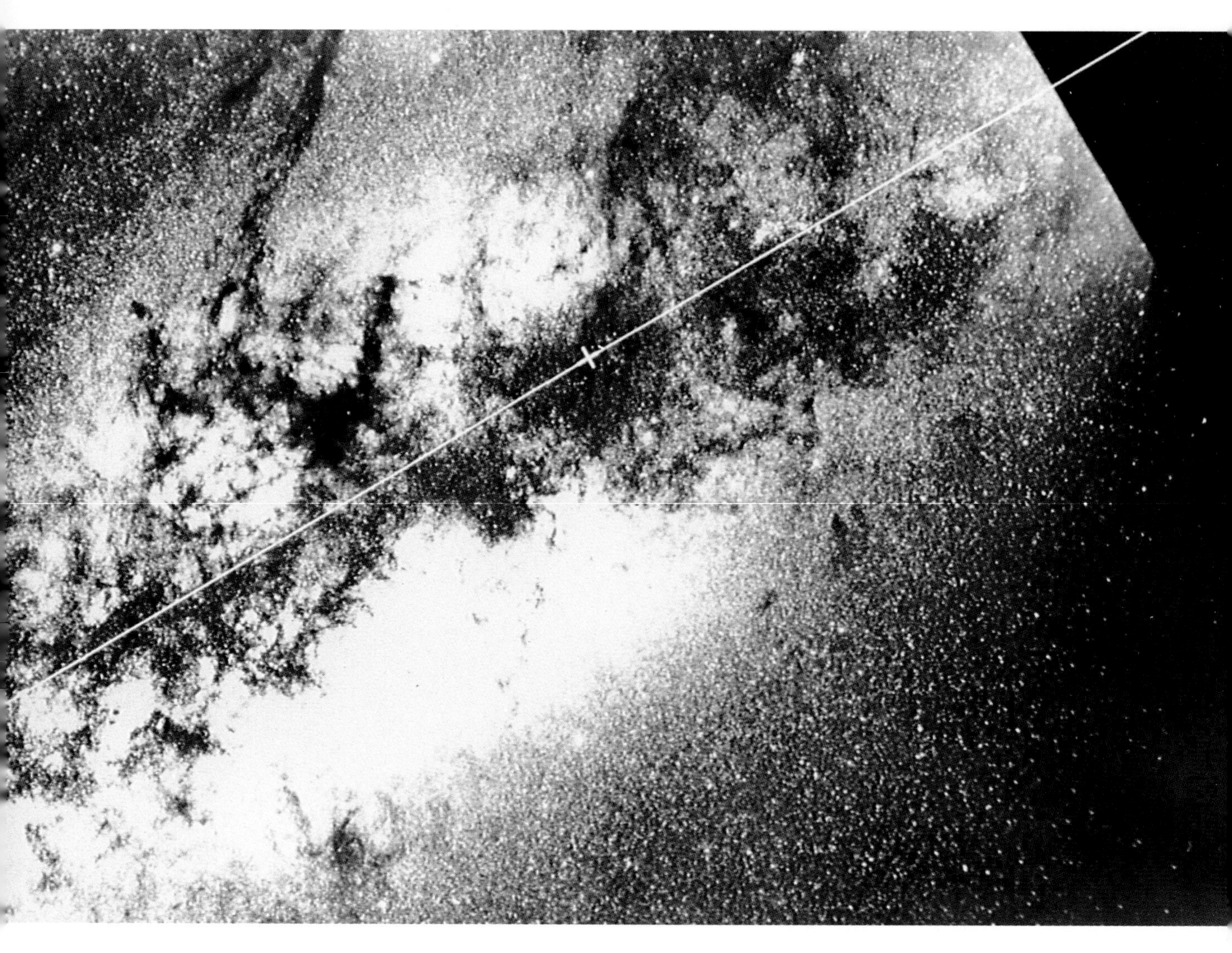

Above: The galactic center, the central Milky Way, is seen here in this photograph by the famous earth-bound telescope Mount Palomar. *(NASA/Mount Palomar)*

Neither massive nor hot enough to become suns, they settled into "cold bodies" wheeling about stars. This was the beginning of planets and moons. The lesser clumps became rocky asteroids and icy comets, orbiting a star under the same gravitational laws as the planets.

An even more miraculous transformation had begun. Bombarded with radiation from nearby stars, with cosmic radiation from throughout the universe, from chunks of solid matter transformed into liquids, as well as from lightning blasted free of immense negative and positive electrical forces, many different types of debris were undergoing chemical reactions of amazing proportions: The most primitive forms of plant and animal life were beginning. These new evolutionary beings were able to reproduce themselves, and on worlds with water and other liquids, with temperatures neither too hot nor too cold, life flourished.

Some four billion years after the Big Bang, trillions of spiral-shaped galaxies formed from inrushing floods of matter and newborn stars.

Among them was our galaxy, the Milky Way.

Seven billion more years passed and another new star came to life—our sun. It collected huge clouds of dust and gas and the ricocheting debris of other stars. All this matter coalesced into planets and moons.

The third planet from this sun would be called Earth.

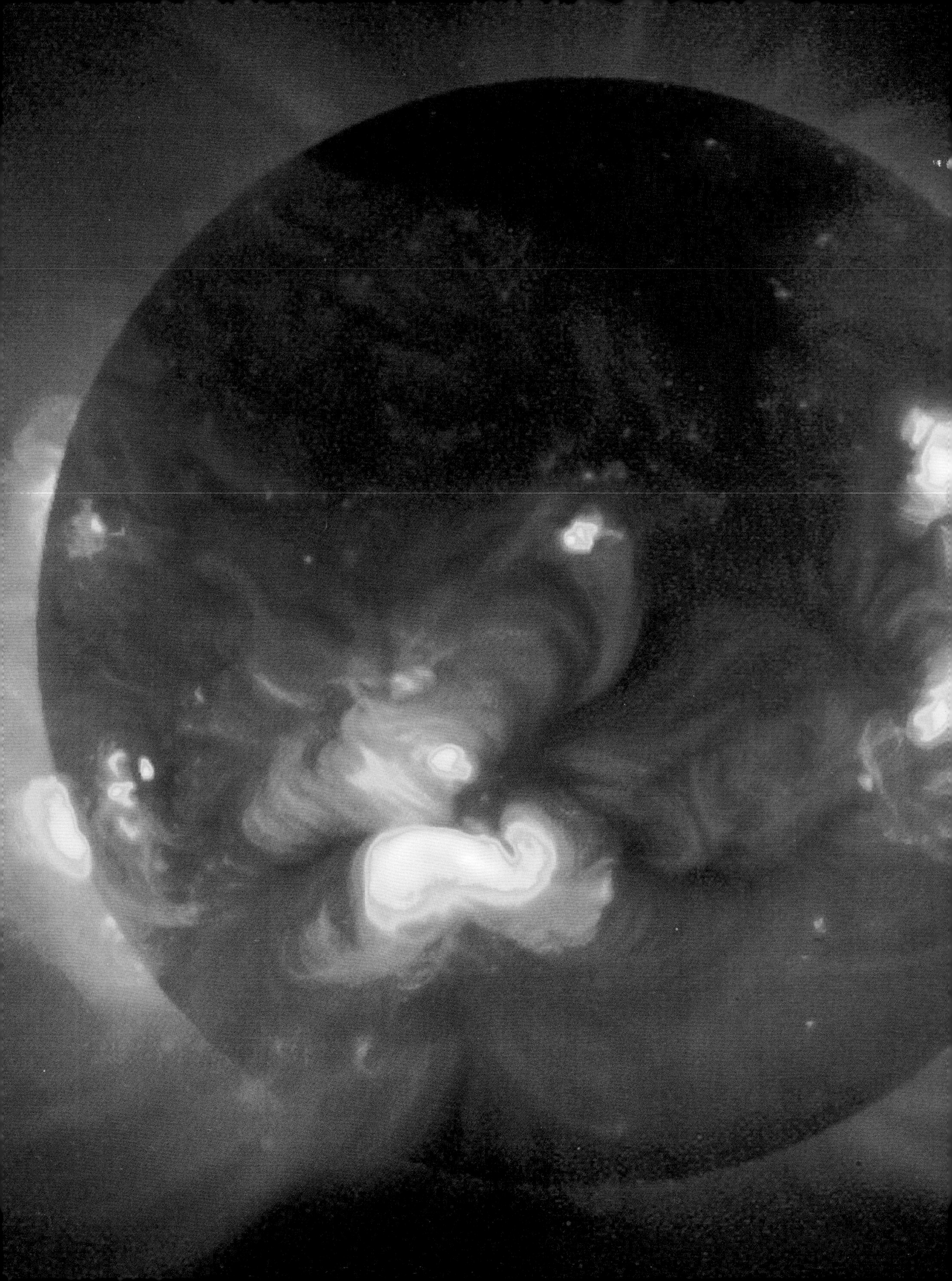

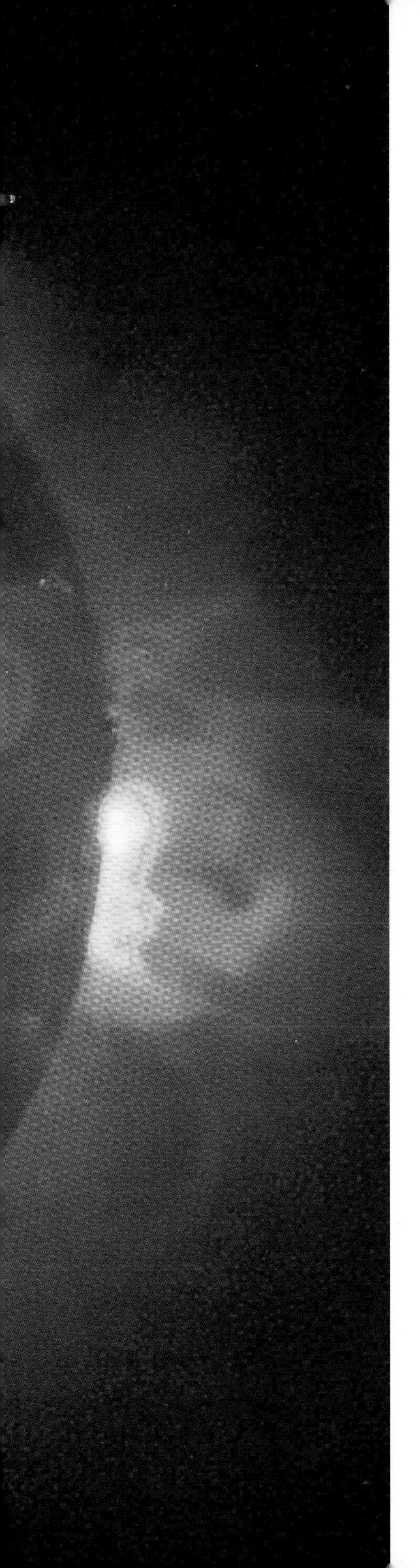

2 SPLENDOR SOLIS

There is one star the Hubble Space Telescope will study only on special occasions, in its search for answers to the mysteries of the universe—a star that in visible light is 864,000 miles in diameter. In the spectrum of radio waves it is twice that size. It is a yellow-type star and has been burning for an estimated 4.6 billion years. Every second, this raging thermonuclear furnace consumes four billion tons of matter. It has been doing that for more than four billion years. As stars go, it is neither large nor small. It's big enough to absorb a million planets the size of Earth, yet other stars are nearly 400 times the diameter of this particular sun.

It lies in one of the outer spiral arms of the Milky Way, thirty thousand light-years from the galactic center. Undistinguished, calm and steady as stars go, it is one of millions of stars in this curving arm of the galaxy. Yet it is the single most important star we know, or ever have known, or ever *will* know.

It is our sun.

Though Hubble will not be used to study it directly, everything about this furnace on which all life on this planet depends is critical to Hubble's task. Our sun is the yardstick by which we mea-

Previous page: This breathtaking image of our sun was taken by Lockheed's Soft X-ray Telescope on board the Yohkoh satellite, which was placed in orbit by a Japanese rocket in August 1991. *(Lockheed)*

Below: Sunrise over the moon.

This color-enhanced image of the sun rising over the moon was taken by the Clementine spacecraft during its 71 days in lunar orbit in early 1994. The planet Venus can be seen to the right of the solar corona. *(Naval Research Laboratory)*

sure the nature and the energy of all other stars. Other suns are smaller or bigger, live and die more quickly, or will burn long after our sun loses its brilliance and begins a long and slow dying. We measure the colors, the brightness, the steadiness, or the pulsations of other stars using our sun as the yardstick. Without our sun as a guide, the study of other stars would be meaningless.

For hundreds of years, astronomers judged our sun as an average star of medium size and density, burning in a sedate and reliable manner. Along most of its surface it blazes with a temperature of 11,000°F. Its interior is another matter, seething furiously at 15 million degrees.

Before the space age sent instruments to study the sun without Earth's atmospheric muck sorely restricting accurate judgment, much of what we accepted as fact about that star was distressingly wrong.

We thought that compared to what Hubble and other telescopes have shown us, the universe was a far-reaching neighborhood of unexpected violence, while *our* sun performed like a maiden aunt.

Above: This is a double star system found by Hubble in the heart of the constellation Sagittarius.

One member of the binary is a neutron star, an extremely dense and compact corpse of a massive star that exploded 28,000 years ago. This image spent those 28,000 years racing across space at the speed of light to reach Hubble's cameras August 27, 1993.

The neutron star is slowly devouring its larger but less-massive white-dwarf star companion, seen at lower left. *(NASA)*

We saw stars imploding and exploding, red giants that could swallow almost our entire solar system, binary systems in a life-and-death struggle of one star devouring its lesser companion. Supernovas detonated in cataclysmic blasts that outshone the combined light of billions of stars. Entire galaxies thundered into one another, destroying trillions of stars. The bursting radiation was so severe it prevented life from forming on any planetary bodies that might be within trillions of miles.

But here in the distant stellar suburbs of our galaxy, our own solar system seemed like classic stability. Planets revolve about our sun in stately procession with intervals measured reliably to thousandths of a second. Moons swing about planets with the same predictable orbits. And even our sun itself seemed to wheel obediently along its gravity-commanded course, orbiting the galactic center like billions of companion stars.

But this benign image proved misleading. Now that we study the sun with instruments orbiting our planet and looping about

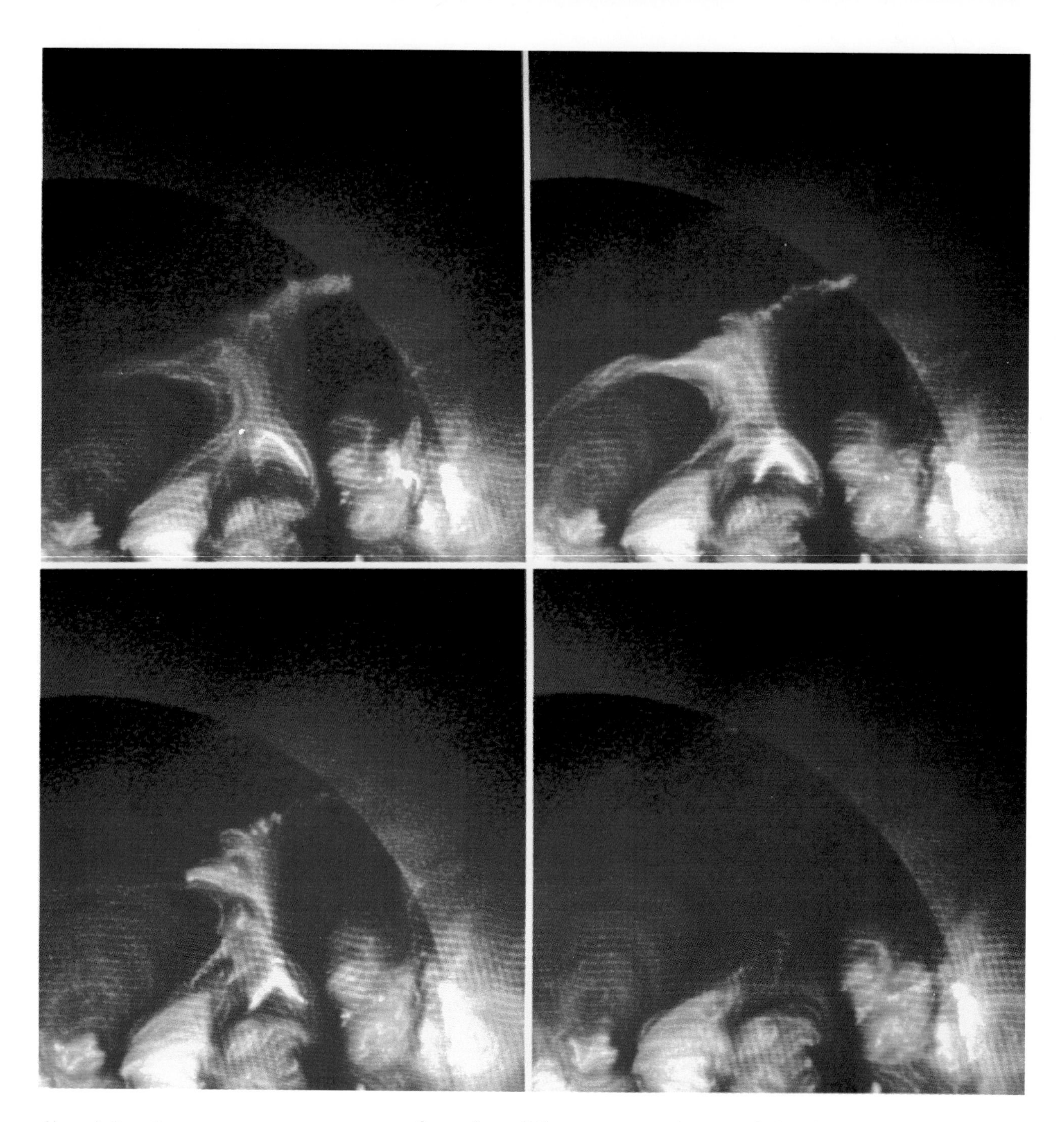

Above: In these four sequence images of the sun's southeast limb, Lockheed's Soft X-ray Telescope on board the Yohkoh solar research spacecraft shows us the rapid restructuring of the sun's magnetic field. *(Lockheed)*

the poles of the sun, our picture of that benign thermonuclear furnace has changed. The sun may be sedate compared to an exploding star, but it is less than quiescent, thrashing within its core and through its atmosphere with mighty spasms, storms, and eruptions.

Hubble may study another star that matches *our* star closely in size, temperature, density, and other characteristics. During such an optical and instrumental excursion, if Hubble finds imbalances in the burning of that star, we may be getting the first warnings of what might cause our own sun to behave in a similar manner. Hubble may give us the extraordinary opportunity to best prepare our world for oncoming calamity.

That alone is worth every dime spent on the Hubble Space Telescope.

There is a strange payback for looking in both directions. Hubble's cameras and instruments record stars in various stages of development. By recording the different growth periods, from the point of matter condensing to the point where a star burns steadily, Hubble gives a look back in time to witness the very same events that created our sun.

Below: To truly understand our own star, scientists must study its activity over a period of time. The images shown here were photographed by Lockheed's Soft X-ray Telescope on board the Yohkoh solar research spacecraft over a ten-month interval.

The photographs show the sun during relatively active and quiet coronal X-ray emissions. *(Lockheed)*

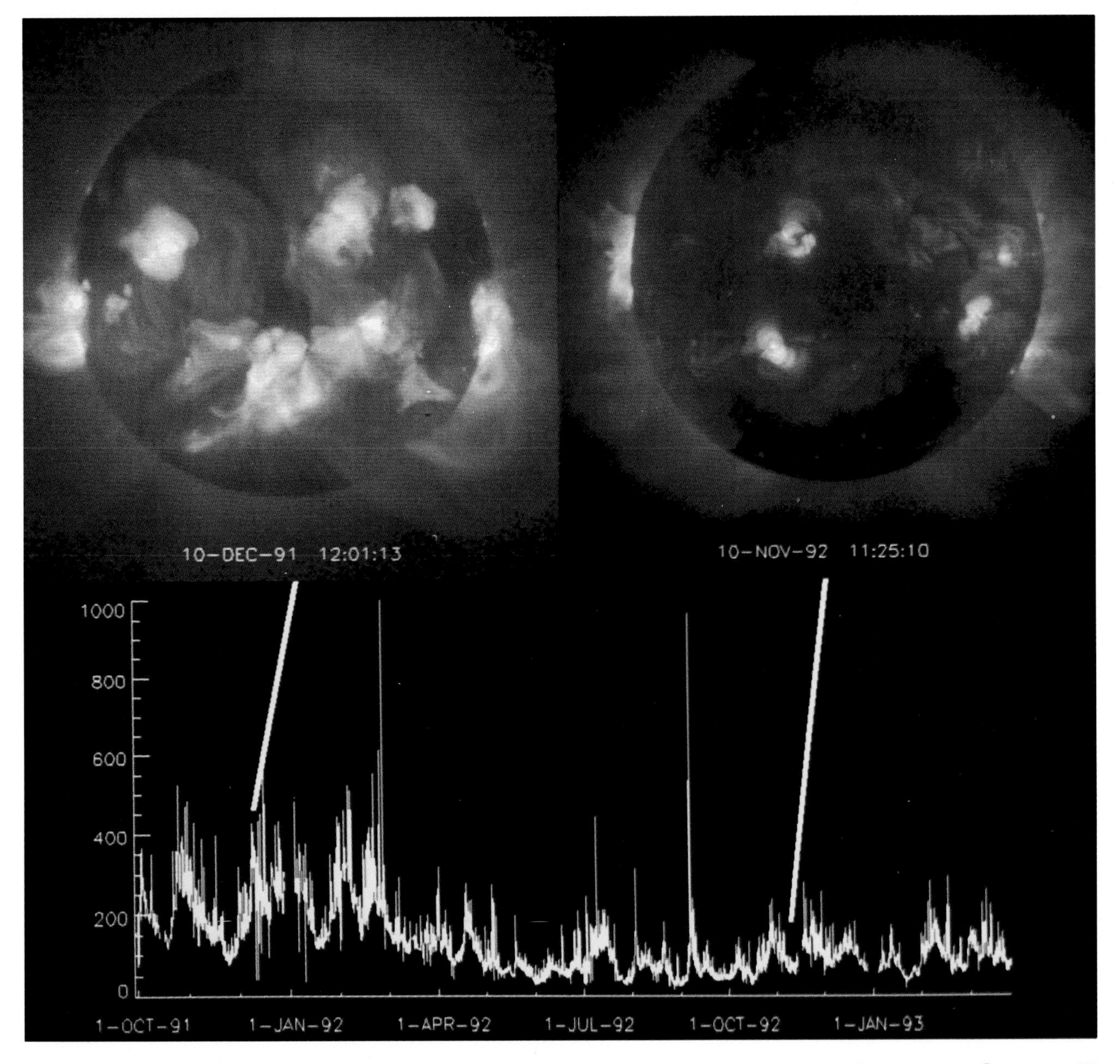

Billions of years ago, what would one day contain our entire solar system was a vast flood of hydrogen and helium, ionized plasma, and the heavier elements spawned from the cores of exploding stars. Borne on the shock waves of titanic stellar blasts, all this drove through a vacuum as the raw material of our future. The stellar debris began to compact, the lightest elements mixing with heavy elements cooked in stars before they exploded.

As they compacted, the clumps exerted increasing gravity on other matter about them. Like moths fluttering to a bright light, elements accelerated toward the clumps. Over billions of years, the process fed upon itself. The greater the mass the clumps accumulated, the greater their gravitational pull and growth. Infalling material began its spiraling motion. The conservation of angular motion kept increasing both spin and mass.

The clumped mass became a sphere, as if a giant had kneaded and shaped raw material into a protostar.

The unborn sun had much greater mass than the star we circle today—perhaps fifty times greater. Hubble still seeks those secrets.

As gravity ground atomic particles together, they became thicker and heavier. There was so much inward pressure the core glowed dark red, then hotter and brighter. Heat rushed outward, a twisting gale exhaled by the still-forming sun. The outward push of heat was counteracted by the inward ram of gravity. And as hydrogen was converted to helium, still more heat and energy were released.

The fire had been lit. The pulsing outward pressure was so great it should have torn apart the protostar. Gravity commanded a balance—raw energy causing expansion, gravitational force causing contraction.

For a million years the struggle continued. The protostar was eight *billion* miles in diameter. Red yielded to yellow-white as internal temperatures increased. The protostar cast away its thin outer shell of gas and shrank steadily to a diameter of 1.7 million miles, just twice the visible-light diameter of the sun we see today.

A huge disk-shaped cloud billions of miles wide had been forming about the star all the while. Irregular clumps and clods gathered within the cloud, each sucking in raw matter like cold vacuum cleaners. Eventually, they compacted into the spherical shapes of planets.

Another forty million years went by. The sun contracted more slowly now; its light was fierce yellow. It had become a sun with planets, all stable. It was still 4.6 billion years in our past.

Left: The outer surfaces of the sun blast in a spectacular display about the solar disk during a total eclipse by the moon.

This magnificent total eclipse was photographed by Mike Hutton, Director of Brevard Community College's Astronaut Memorial Planetarium, from the deck of the MS *Jubilee* cruise ship 20 miles off the coast of Mazatlán, Mexico, July 11, 1991. *(Mike Hutton)*

Right: This full-disk image was taken by Lockheed's Soft X-ray Telescope on the Japanese Yohkoh orbiting spacecraft. *(Lockheed)*

As they would say in a yet-distant future, on a small dense water planet third from this sun:

Splendor Solis. A star is born.

Compared to its planets, the sun is overpowering in size and fury. More than 99 percent of all the solar system's mass lies within our star. And like planets and moons (and other stars), our great ball of fire rotates on its axis. It is a quirky movement. At its equator, the sun rotates one full turn every 25 earth days, as does most of the upper 30 percent of the sphere. But at the latitudes in between, a full rotation requires 27 days. At the polar regions, the trip requires 35 days. It's clear to astronomers that the wrenching differences in rotation produce violent storms within the great fireball.

Why does this massive burning globe whirl with such variation? When Ken Libbrecht of Cal Tech asked his associates that question, they answered by throwing up their hands. It is a feature that Hubble will attempt to confirm with the stars closest to us beyond the sun, which also have mystifying rotations.

The spectacular outer surfaces of the sun—the chromosphere, corona, and prominences—are dazzling white striations that writhe and blast upward about the solar disk. They can be most easily seen during a total eclipse of the sun by the moon, when the moon blacks out the solar body but leaves visible its spectacular royal crown of stardom. Total solar eclipses come once every one and a half years. If necessary, Hubble can erect a disk before its camera to black out the solar body, photograph the outer atmosphere, and perhaps gain information that will explain how the corona can blaze millions of degrees hotter *above* the sun, when the surface temperature is a comparative icebox at only 10,000 degrees.

Lockheed space scientist Keith Strong said, "That's very strange. You can't heat an atmosphere to several million degrees with something that's only a few thousands of degrees."

No one has told the sun it can't be done; the mystery is still with us.

Astronomers—and Hubble scientists especially—would dearly love to understand more about our local star, which they once believed they knew so intimately. And mysteries that are confusing at our doorstep become monumental light-years distant.

After many years of solar research, John Harvey, of Kitt National Observatory in Arizona, admits, "We are incredibly ignorant

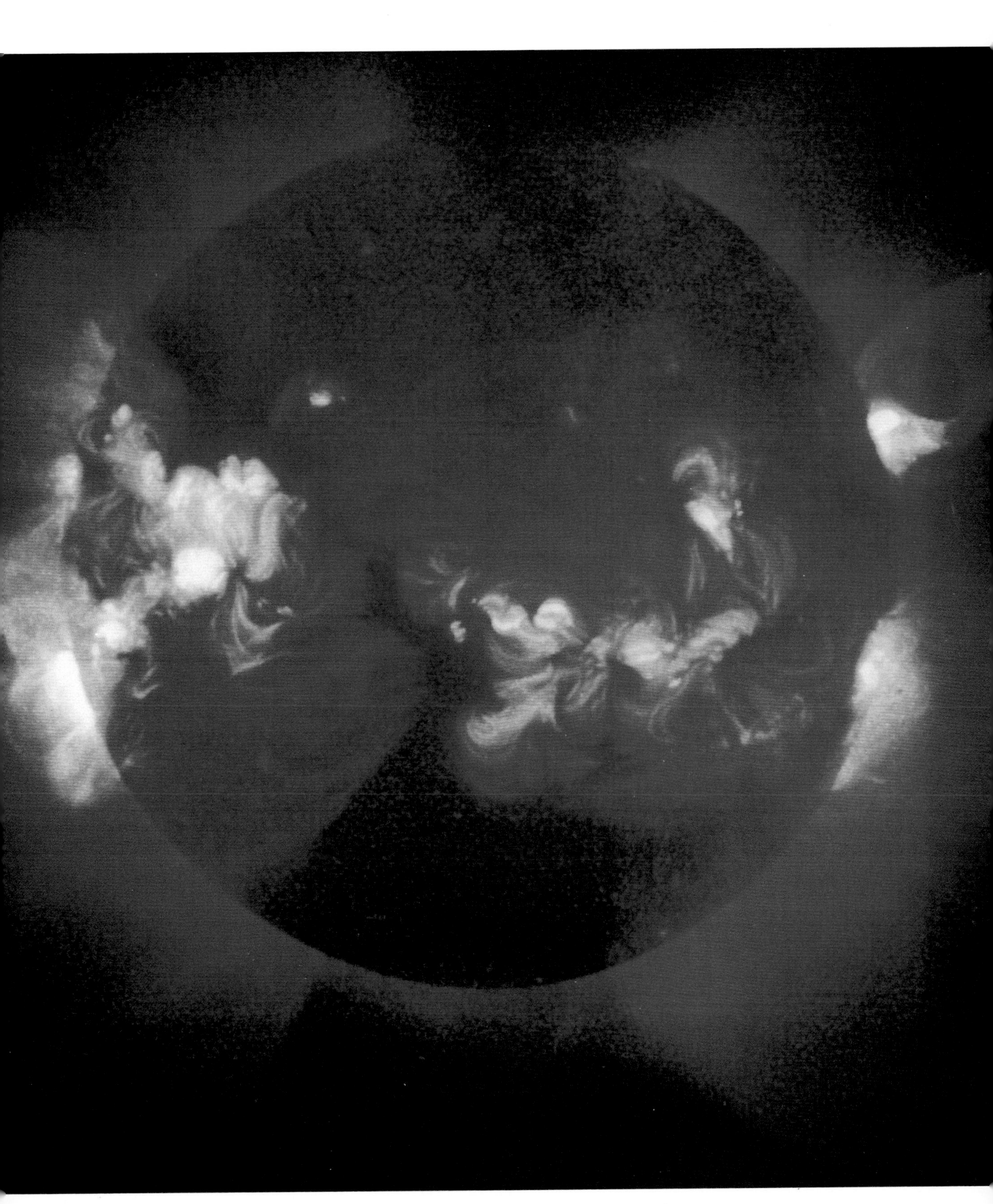

about what's driving solar activity." Violent eruptions, spasms, shudders, and huge "cold spots" form like dark canyons on the surface; we don't know what causes these effects. The energy release that rips outward for trillions of miles begins deep in the solar core and crawls and grinds its twisting way to the surface in an upward struggle that takes 20 million years to burst free of the sun.

At periods of eleven years, sunspots, cooler than surrounding areas and covering several million square miles, appear suddenly on the surface, twisting to hurl forth tremendous magnetic fields. This chokes off normal heat and creates "cool spots" that act as solar spark plugs for explosive eruptions along the sun's surface. Monstrous prominences, spears and whorls, and curving arches, all made of solar gases, stab hundreds of thousands of miles into space. Coronal discharges leave the sun as stupendous volcanic eruptions, hurling enormous spears of powerful subatomic particles in every direction.

The sun writhes with enormous magnetic fields. Its different rotating speeds twist the lines of magnetic force like stellar taffy. When the magnetic fields break free, great clouds of particles cascade outward. On Earth, we become aware of the eruptions when radios crackle with static, power stations break down, power lines and transformers explode, television reception is snowy. What is merely inconvenient here on earth becomes a deadly rain in vacuum for orbiting astronauts and cosmonauts. Scientists have given us detailed maps of solar magnetic fields, especially of the north and south magnetic poles. For decades we believed that the magnetic polar caps reversed positions. It was a solar procedure about which we've been positive for decades.

Except that we've been dead wrong. In the early fall of 1994, NASA's Ulysses solar probe looped about the southern hemisphere of the sun and passed almost directly beneath the south pole. At last, the opportunity was at hand to obtain detailed, close-look data on the characteristics of that polar region.

Alas—there isn't any magnetic south pole! Because of the superbly functioning Ulysses, a familiar refrain sounded here on Earth. "We'll have to rewrite the astronomy books—again."

The distance between Earth and sun rounds out to 93 million miles, or about eight minutes at the speed of light.

The sun is many things to many people and for different reasons, but it is life and existence for *all* of us. Since the Hubble Space Telescope slipped into orbit in 1990, and was later repaired by the crew of Endeavour, the sun's new meaning as a yardstick for studying distant stars has been in full use. If Hubble finds other middle-aged, long-lived, medium-sized yellow stars—like ours—the search will be intensified.

That's where we just might meet our first extraterrestrials. Friendly aliens.

We hope.

3 INNER CIRCLE

As the Hubble Space Telescope team intensifies its search for planets orbiting stars other than our sun, it will be well prepared to measure what's out there, using as yardsticks neighboring worlds much closer to home. If the team finds anything like the two planets closest to our sun, it will quickly check them off as worlds that failed abjectly to become habitats for life. The possible alien equivalents of Mercury and Venus won't even fit the common saying "It's a nice place to visit, but I sure wouldn't want to live there."

But such comparisons are vital, for the Hubble team is primed to search for planets *with life*. Anything like Mercury would be dismissed out of hand.

Mercury is a misbegotten runt of a planet, racing around the sun at nearly 30 miles a second—a furiously paced orbit that averages 36 million miles from the sun. Needless to say, as the nearest planet to the sun, Mercury is seared, baked, scarred, and lashed by heat great enough to set everything on our own world ablaze, save hard rock and harder metals.

To planetary scientists, and especially biologists, Mercury is a

Previous page: *This global view of the surface of Venus was taken by the planetary craft Magellan. (NASA)*

hunk of spherical rock, interesting only because it is there. The tiny planet, its diameter of 3,100 miles making it the smallest of nine worlds circling the sun, is gravid with heavy iron in its structure. But not even overdoses of heavy elements could retain any atmosphere on Mercury. The unremitting blast of solar energy billions of years ago whipped away whatever primeval atmosphere might have collected during the time the planet assembled from debris and flotsam dragged in by immense solar gravity.

Even the planetary motions are a nightmare. Mercury rotates on its axis once every 59 days. It takes 88 days to complete a revolution about the sun. That's earth days; Mercury's own days and nights only make for confusion. The lifeless world rotates on its axis three times for every two revolutions around the sun, exposing its surfaces to a steady pounding of heat and hard nuclear radiations.

Below: NASA spacecraft took these six images of planets in our solar system, including Earth's rise over the moon's surface with the sun flare on the edge of the Earth's limb. The first planet above the moon is Venus; top left to right are the planets Jupiter, Mercury, Mars, and Saturn. *(NASA)*

Above: This is the planet Mercury, the nearest planet to the sun. The picture leaves no question that Mercury's surface is scorched by intense solar heat. This mosaic of Mercury was photographed 124,000 miles away by the Mariner 10 spacecraft as it approached the planet March 29, 1974. *(NASA)*

Imagining what it would be like to stand on Mercury isn't very difficult. The pressure suits enabling astronauts to walk and drive about the lunar surface wouldn't last very long on the hellish surface of Mercury—just enough to let whoever was so foolish to chance that excursion to die badly and quickly. On the sunlight side of Mercury, nuclear radiation would rip through suit and astronaut with much the same effect as standing naked before an atomic bomb explosion. A surface temperature of 950°F would quickly turn the interior of the suit to boiling steam—most of it from the astronaut's own body liquids. Likely the Mercury walker would be dead well before internal heat-generated pressure became so great the suit exploded.

Staying on the dark side, keeping pace with rotational speed to remain in permanent shadow, would be only a stopgap measure. In the utter blackness of Mercury's craters, the temperature plunges to minus 346°F; it would take a lot of suit power to keep an astronaut healthy, breathing, and mobile. And should the space travelers delay remaining in shadow, even a brief stumble into the omnipresent solar barrage would bring down the curtain.

In many ways Mercury is eerily like our moon, battered with craters huge and small, pocked with dust, and marked with basins almost impossible to distinguish from the lunar surface. Enormous cliffs crisscross the planet, some more than a mile above local terrain that stretches for a thousand miles. The single most dominant planetary feature is the Caloris Basin, its interior filled with ancient lava. And like our moon, there are clear signs of enormous upheavals of the planetary crust following massive meteor

Above: As Mariner 10 approached Mercury at nearly 7 miles per second on March 29, 1974, it took this picture from an altitude of 21,700 miles.

This photograph shows a heavily cratered surface with many low hills. The large valley to the right is 4.5 miles wide and 62 miles long. The large, flat-floored crater near the center is about 50 miles in diameter. *(NASA)*

impacts. Much of Mercury's terrain is so much like the Imbrium Basin of our moon that photographs of both worlds are almost impossible to tell one from the other.

There is one interesting feature of the dead world worth mentioning. An observer standing on Mercury (or sealed within massive walls of armor and glass to survive penetrating thermal radiation) would witness a sun gone berserk. The sun moves eastward through the airless black sky, and then due to Mercury's odd mixture of rotation and revolution, comes to a halt, hanging motionless before the observer. Soon it would again move, but this time it would slide backward for the next eight Earth days.

On our home world, we're accustomed to high rotational speeds. Standing on the Earth's equator, we know we're spinning eastward at a thousand miles per hour. Mercury would knock our biological clocks askew; it turns at a turgid, befuddling crawl of only six miles per hour.

As a yardstick for potential life, Mercury's message to the Hubble team is clear: *Move on.*

FALSE GODDESS

Two planets could hardly be more similar in physical size than Venus and Earth, almost as if they were poured from the same mold. At 7,600 miles in diameter, Venus is only slightly smaller than Earth. For example, a man who weighs 200 pounds on Earth would weigh 174 pounds (in earth numbers) on Venus.

In the night sky, both worlds gleam with brilliantly reflected sunlight. Both are heavily clouded, both are part of the "inner circle" of four small, dense worlds orbiting close to the sun. Besides the moon, Venus is the glistening wonder in the night sky, a beacon of startling silver-white six times brighter than Jupiter and fifteen times brighter than the brightest star, Sirius. Small wonder the ancient Romans named this sparkling planet after their goddess of love and beauty.

When the two planets are at their closest, Venus is but 26 million miles away. It has always been a diamond gleaming in darkness, but the similiarities disguised Venus' true nature—this second world from the sun is a false goddess.

Astronomers first thought Venus was a planet soaked with torrential downpours, lush with vegetation, its jungles and islands

Below: The single most dominant feature on the planet Mercury is the Caloris Basin, seen in this photograph taken from a range of 11,800 miles by Mariner 10. The Caloris Basin's interior is filled with ancient lava, and, as on our moon, there are clear signs of enormous upheavals of the planetary crust following massive meteor impacts.

Mercury's terrain is so much like that of Earth's moon, photographs of both worlds are at times almost indistinguishable. *(NASA)*

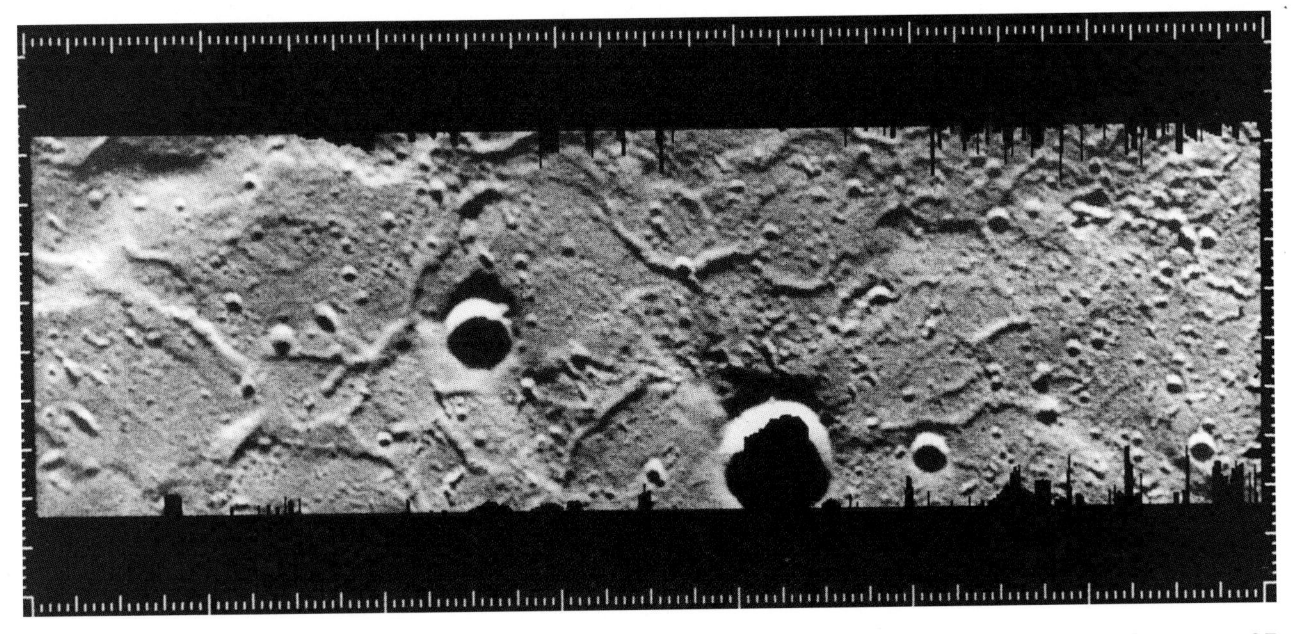

packed with dinosaurlike creatures, all a mirror of ancient Earth. Certainly it was warmer than our world due to its closer proximity to the sun, and it may once have had the unique ability to bear both monster lizards and a civilization of men and women much like us.

Nineteenth-century astronomers were sure Venus was a delightful world, with millions of farmers living beneath the friendly hand of benevolent emperors. The renowned German astronomer Franz von Paula Gruithuisen pointed out a mysterious faint glow on the dark side of the planet. His associates believed the glow came from reflected earthlight, just as the moon is awash with light reflected from our planet. Gruithuisen declared this was impossible; No moon orbits Venus, thus there can be no reflected moonlight.

Possible or not, astronomers worldwide confirmed the night glow on Venus through telescopes. Gruithuisen was thus able to

Below: This picture of Venus with its mysterious glow was taken by the Galileo spacecraft. *(NASA)*

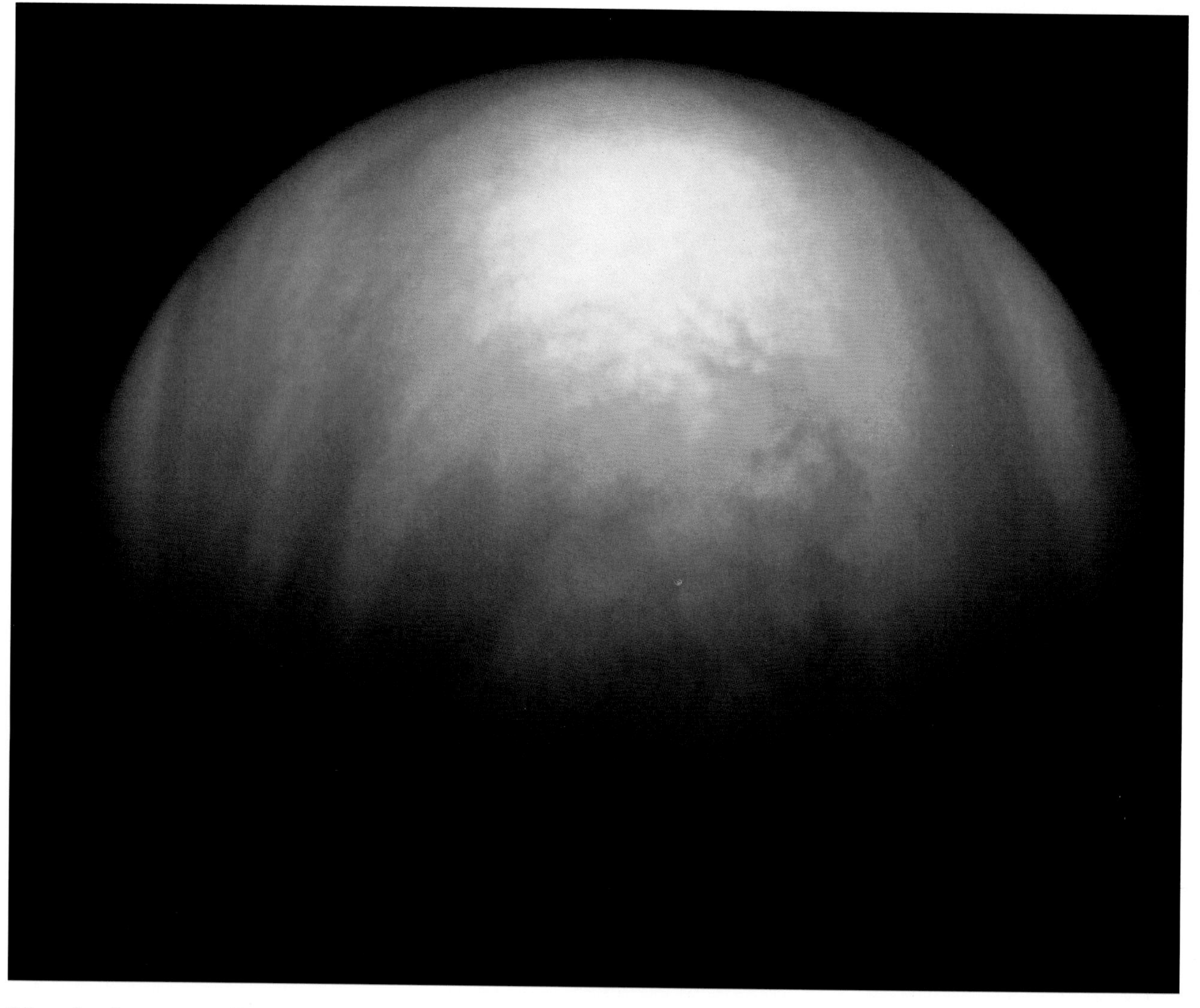

"prove" a Venusian civilization whose inhabitants enjoyed a lifetime of 80 earth years—equal to 130 Venusian years. That ghostly gleam in the dark came when a new emperor was being crowned and millions of people held a torchlight festival in his honor.

No nineteenth-century astronomer had any way of knowing it was violent, massive lightning bolts casting the glow.

Venus remained the bright, sparkling mystery in our evening and morning skies well past the halfway mark of the twentieth century. December 1962 was when the first planetary probe, Mariner 2, sped past the planet within 21,600 miles of its cloud tops. That's when we began to shred the fantasy.

American and Russian probes orbited Venus, dropped scientific stations onto its surface, and sent other research apparatus drifting beneath balloons along the Venusian clouds. Russian landers stunned scientists when they returned true-color photographs

Above: Magellan, the extraordinary American spacecraft, took tens of thousands of pictures with short, powerful bursts of radar energy that pierced Venus' thick, poisonous clouds. They revealed the hellish world for what it was.

This view of the surface of Venus shows lava flows that extend for hundreds of miles across the planet's fractured plains. *(NASA)*

Above: The Venus we see today, although it formed the same time as Earth, 4.6 billion years ago, is simply a "new" planet. No scientist knows why it underwent this cataclysmic change. Scientists only know they are reluctant to speculate. *(NASA)*

of the surface, showing flattened and sharp-edged rocks, a world of dull orange-brown color, thick haze, and dusty soil everywhere.

All these missions provided us with views of bits and pieces of the planet. Not until the extraordinary American orbiter Magellan took tens of thousands of pictures—with short, powerful bursts of radar energy that pierced through clouds as if they didn't exist—did we see the hellish world for what it was.

Magellan plunged to a friction-ignited fiery demise in October 1994 after five years of stunning performance. It told of a cataclysmic change about 500 million years ago that covered the entire planet with new surface material and hid almost everything that existed. Although it formed the same time as Earth, 4.6 billion years ago, the Venus we see today is a "new" planet. No scientists know why it underwent this cataclysmic change. They only know they are reluctant to speculate.

But there is no question Venus suffered a disastrous greenhouse effect condemning the world to torturous heat and crushing pressure. Thick clouds rose up to 62 miles, mostly carbon dioxide in the lower depths and poisonous sulfuric acid at higher levels. Infrared radiation passed through this mass in sufficient strength to keep temperature high, increased by the clouds that prevent the heat from radiating back out into space. With no place to go, the heat grew until today Venus endures surface temperatures upwards of 900°F, easily melting lead and zinc and making putty of even denser materials.

Atmospheric pressure at the surface is equally a nightmare. The air is so thick and viscous, the pressure is equal to that several

Below: This false-color image shows the 250-mile-wide volcano Sapas Mons. It rises almost a mile high above the surface of Venus and is flanked by numerous overlapping lava flows. The dark flows on the lower right are thought to be smoother than the brighter flows near the central part of the volcano. *(NASA)*

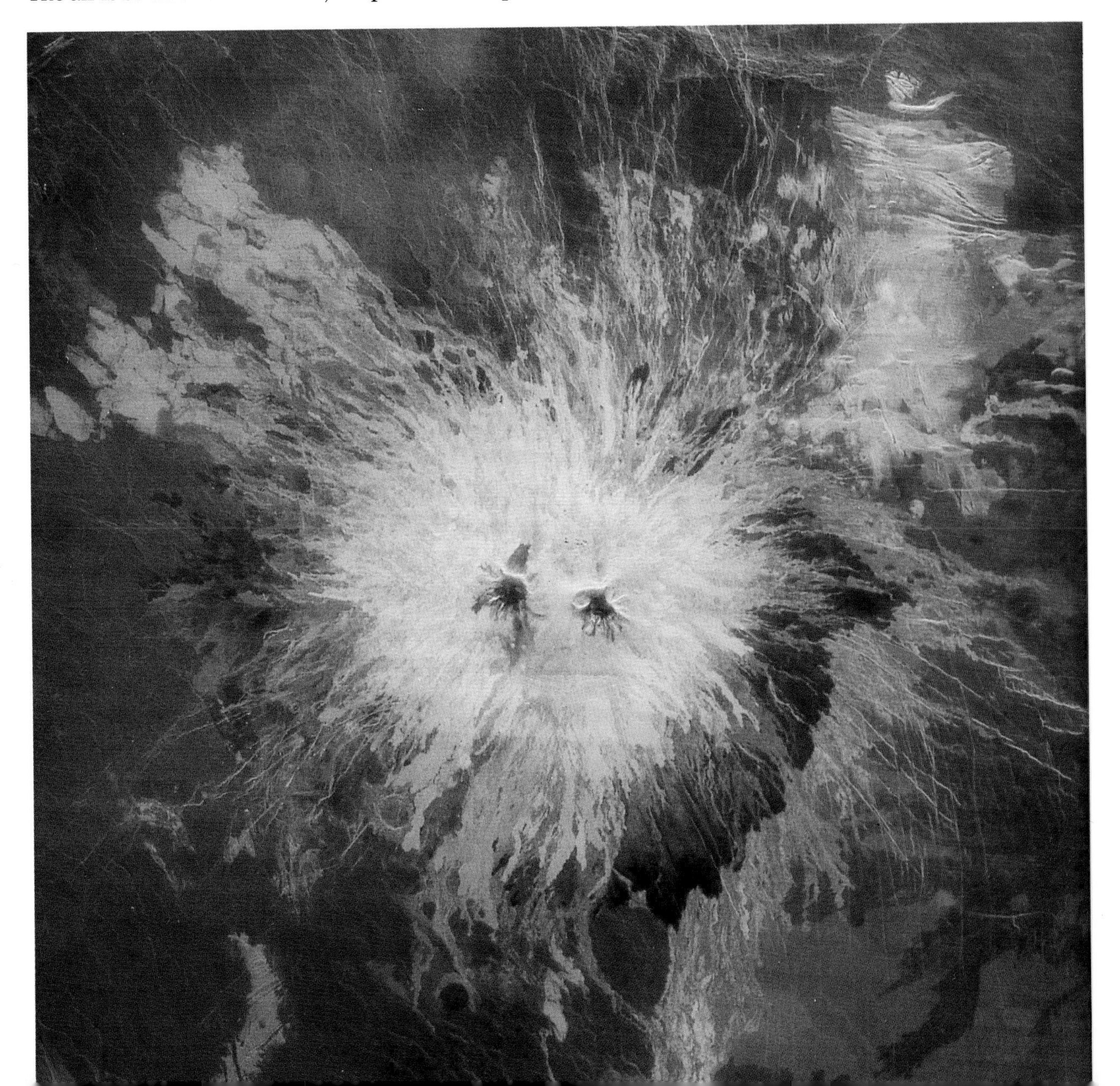

thousand feet beneath the surface of our oceans. At a minimum that pressure is approximately one hundred times greater than Earth's atmosphere, and at times instruments that settled to the broiling surface recorded pressure as high as 3,000 pounds for every square inch of ground.

In all these respects, the conditions of Venus are critical to the Hubble astronomy team. Encountering other planets from afar, the Hubble scientists know that visible-light photographs tell only a partial picture. Instruments that measure energy in infrared and ultraviolet may detect Venus-like planets that invite us with bright clouds but in reality are more like Hades than some pipe-dream Eden.

This is the science of "failed worlds" that has become part of the program planning in the search for planets beyond our own so-

Below: In this image the Gula Mons volcano stands nearly two miles high on the left horizon. *(NASA)*

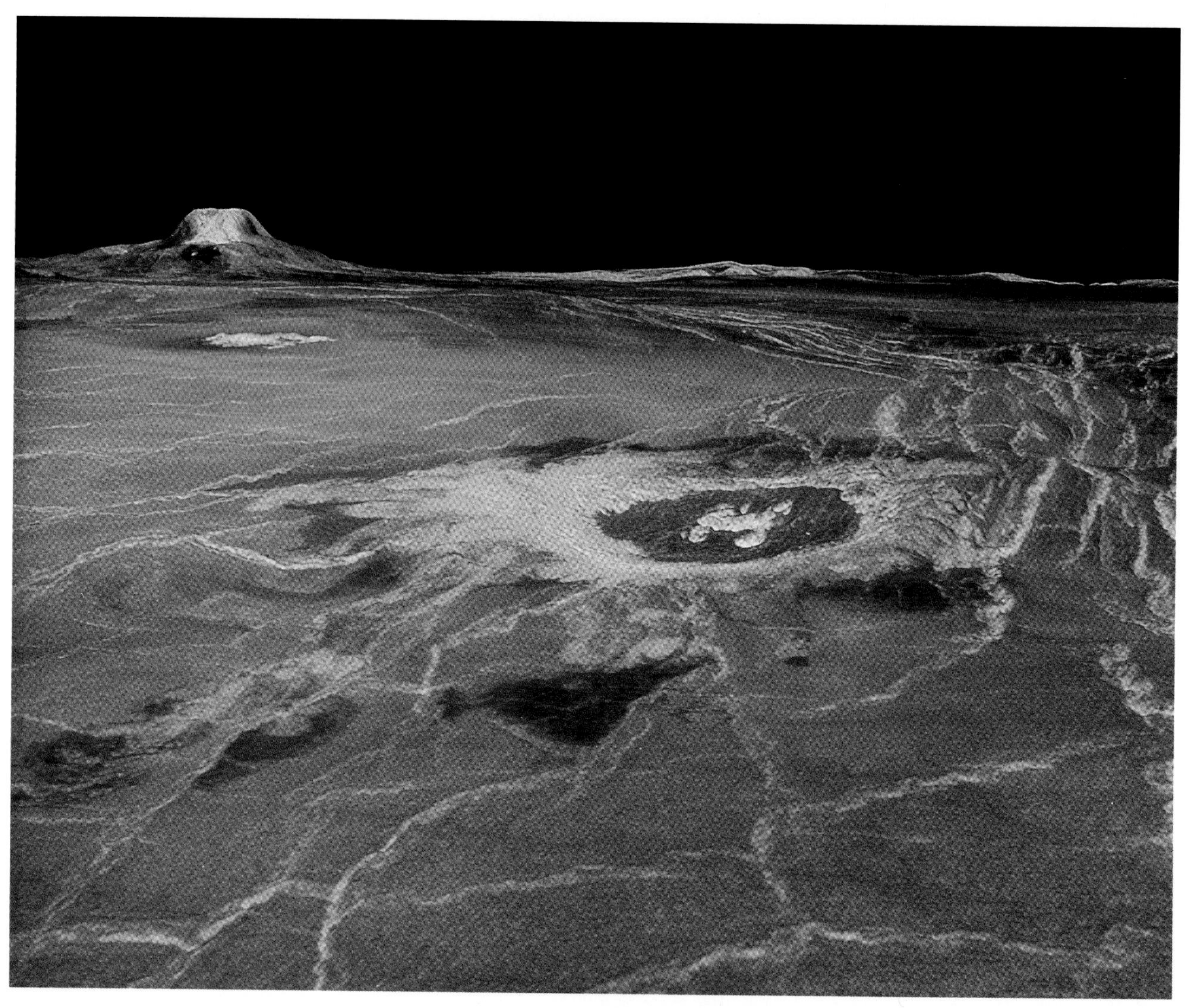

lar system. Even if we were capable of super technology, simply erasing the clouds that produce the Venusian greenhouse effect won't promise a new world for human settlers. Only 67 million miles from the sun, Venus receives much more thermal radiation than Earth—as well as dangerous bombardment by subatomic particles—and would remain intolerably hot for human life.

There lies yet another reason why Venus is a false goddess. Much like Mercury, Venus is a renegade planet as it circles the sun once every 225 Earth days. Venus has an astonishing motion as it rotates on its axis, turning into the west—exactly the opposite of Earth. That's interesting but hardly injurious to life. But incredibly, a year on Venus is shorter than a Venusian day. It takes 243 Earth days to complete just one Venus day. That is condemning to all known biological forms; there is virtually no way to escape the buildup of lethal temperatures.

Below: Maat Mons is displayed in this three-dimensional perspective view of the surface of Venus. *(NASA)*

Above: Weathered volcano domes called "volcanic pancakes" lie along the eastern rim of the Alpha Regio highlands. They have been measured at 16 miles in diameter and a half mile above surface level. *(NASA)*

When the first probes descended by parachute to Venus' surface, they indicated a world of bland ground features, as if temperature and pressure had mashed the planet to featureless putty. Great areas of Venus are just like that, appearing as what a Magellan scientist called a "fairly flat rolling plain." David Anderson, one of the Magellan team, finds most of Venus "bland." It's been likened to Nebraska because, he says, "it seems to go on forever."

But it would be a mistake to judge Earth by the desolate, blinding-white flatness of the Bonneville Salt Flats in Utah; where it is so flat that a man can see telephone poles disappearing over the curve of the horizon. If that same man flies just minutes away, though, he faces craggy high mountains—as if he just flew onto another planet.

So it is with Venus, where Nebraska-flat blandness yields to the Ishtar Terra highlands. Here rears the greatest of all Venusian upwellings—Maxwell Montes, a soaring mountain peak 39,000 feet high, 10,000 feet higher than Mt. Everest.

There are between 800 and 1,000 large craters pockmarking this brutish world. On the flanks of Maxwell Montes is the huge crater Cleopatra, a double-ringed feature similar to such double-ringed formations on Earth, Mercury, and Mars. At 60 miles wide and nearly two miles deeper than the surrounding terrain, Cleopatra is the evidence of a catastrophic meteor impact.

But nowhere to be found are the many thousands of small craters like those on our moon and Mercury. Small meteors do not become meteorites simply because they lack the energy to punch through the muddy atmosphere on Venus. They ignite quickly and burn fiercely and often produce a geological feature known as a *splotch.* This is the scientific term describing a meteor that is not dense enough to survive through the atmosphere, descends to a low level, then explodes like a great hydrogen bomb. No crater results from these blasts, but the shock waves pound the ground into a depressed, rounded, unusually smooth surface.

The winds of Venus defy logic. With such fierce temperatures and massive pressure, scientists were certain the face of Venus would be a constant gale-lashed fury. Instead, the wind hardly reaches what would be a mild breeze on Earth: three to five miles an hour, barely enough to exert any disturbing effect along the surface.

Yet there are rippled dunes on this world, which certainly seems impossible without more than a sigh of wind. In Venus' murky past, scientists postulate, huge explosions of meteors and small mountains plunging down from space hurled tremendous shock waves and trailing winds, piling up the basaltic dirt and volcanic ash into "sand dunes."

Four-fifths of Venus is plastered with sheets of lava, volcanic domes, and upwellings in a fractured, hardened surface. More than 430 big volcanoes—each more than 12 miles in diameter—

Below: Sif Mons, a volcano with a diameter of 186 miles, looms in the background. Its lava flows reach for hundreds of miles to the south; few features so fascinated scientists as the huge spidery lines and shapes on the right. *(NASA)*

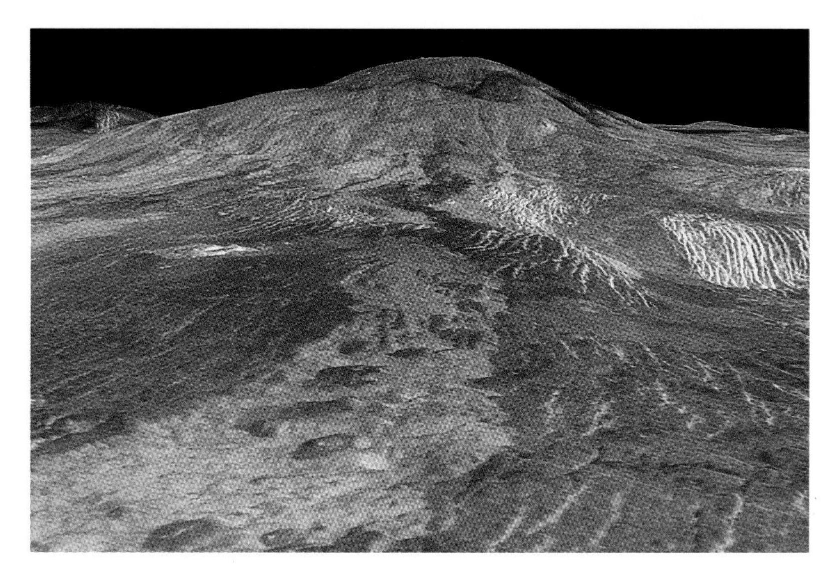

Right: This global view of the surface of Venus from Magellan's synthetic aperture radar suggests the planet is much like Mercury. But when it comes to rotating on its axis, Venus surprises by turning into the west—opposite Earth's rotation. *(NASA)*

loom everywhere on the planet. Tens of thousands of smaller volcanoes were captured by the probing radar of Magellan, but what remained unseen were tens of thousands more engulfed by flowing lava from bigger volcanoes or asteroid impacts that melted the surface and splattered fiery rock in all directions.

Russian landers and parachute-swinging stations found volcanic domes almost everywhere as they scanned the Venusian landscape. They found massive lava flows, impact craters, rills, and ridges, and then, in a world mostly thought of as "bland," a totally unexpected deep canyon that could swallow the Grand Canyon of the American southwest.

Unusual "volcanic pancakes" stretch along the eastern rim of the Alpha Regio highlands, lifting from the surface as flattened, cracked domes. Magellan scientists measured these pancake domes at 16 miles in diameter and a half mile higher than ground level. They would have remained a mystery except that volcanic domes seemed strangely familiar to California-based astronomers, who explained these domes also exist on Earth, especially in California, where they have been weathered into rolling hills and covered with grasslands and forests.

Few features so fascinated scientists as those NASA calls the arachnoids, huge spidery lines and shapes on the planet's surface. Hundreds of thousands of square miles were etched with spidery trails that hint of ancient lava upwellings and sudden, unexpected solidification that remains unchanged to this day.

The most unlikely feature of the planet is the River Styx, a channel gouged in the surface, average 3,000 feet in width—but stretching for an incredible length of 4,800 miles. This is comparable to a single human hair more than a mile long. Stephen Saunders of the Magellan team judges this might have been caused by liquid sulfur or "some other weird compound" that nobody understands.

In 1985, Russian Vega 1 and 2 probes skimmed along the Venusian atmosphere, heated to glowing metal. Forty miles up, the probes separated in two, releasing instrument sections to descend by parachute. The other half from the shell also dropped—slowed by parachute until explosive charges blew away the shroud lines. On each package a balloon shot upward, filled with helium, and lifted javelin-shaped instrument sensors, suspended by a long line beneath the balloon, to begin swift passage high over the cloud-shrouded world. At hundreds of miles per hour they transmitted data on the atmosphere and its deadly chemicals. After 48 hours, their batteries gave out.

Far beneath the lifeless robot spacecraft, Venus boomed and echoed with thunderous blasts of lightning. As many as 25 shattering bolts *per second* pounded the lifeless world, unheard, sending faintly flickering light upward through the clouds—the mysterious faint glow on the dark side of the planet seen by nineteenth-century astronomers. In the thick atmosphere, the pressure waves and thunder were like constant atomic explosions. One blast recorded during the brief lifespan of a Russian station on the surface sent out crushing shock waves that boomed nonstop for 15 minutes.

Unseen and unheard, the violence continues. It is, for a long time to come, the end of a story, except for what Venus warns us may lie in wait out among other stars.

These two worlds, Mercury and Venus, are without life, without flora, without moons.

From now on our gaze, our dreams, our telescopes, and our spacecraft point outward—to Mars and beyond.

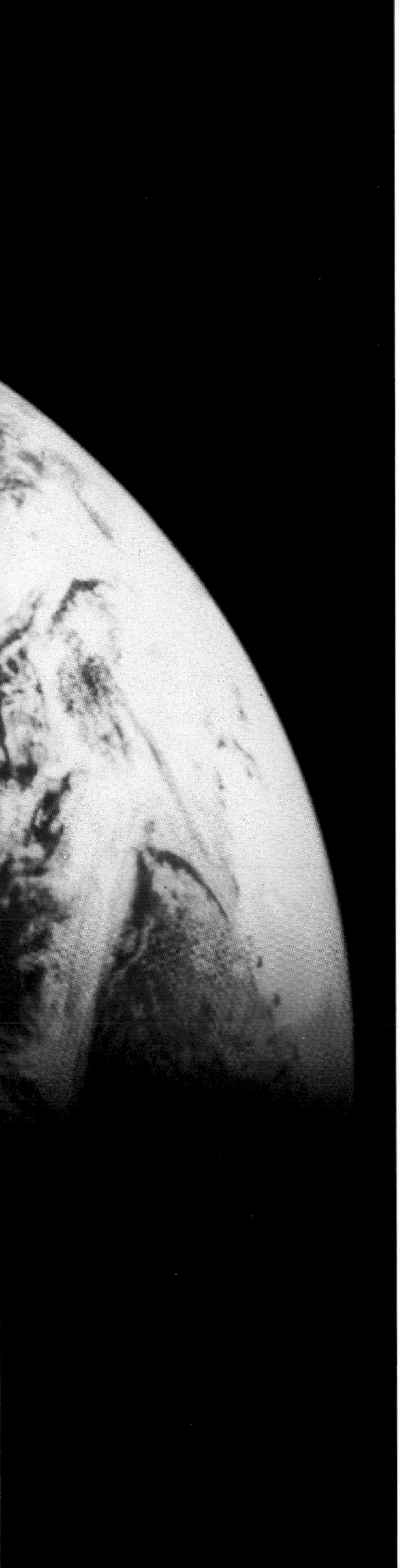

4 WATER WORLD

We launched the Hubble Space Telescope into orbit to survey the heavens within and beyond our solar system. Its mission is to learn more of everything within our reach of optical and instrumented systems, and since it has been repaired, its work has been stunningly successful.

How ironic, then, that every planetary body Hubble has observed and measured is the antithesis of the world from which it rose.

If Hubble discovers one type of planet in particular, one among thousands or even millions, a water world, orbiting a star with all the requisites to enhance the start and reproduction of life-forms, that discovery will mark a new age of judging life in the universe. With that discovery will come the conviction that the universe has been seeded with planets that produce intelligent and creative life-forms.

To date we know of only one such planet. It is blessed with a weather engine that produces seasons that are kind to life-forms, it turns at a rate superb for creating life-sustaining cycles of light and darkness. It is surrounded by a captive and incredibly thin at-

Previous page: This is a picture of our water planet from 1.2 million miles in space. It was taken by the Galileo spacecraft on December 11, 1992, as it left the planet on its 3-year flight to Jupiter. Antarctica is visible at the right of the image, and dawn is rising over the Pacific Ocean. *(NASA)*

Right: This fantastic picture of Earth was taken by the Apollo 17 astronauts during their journey to the moon in December 1972. The outstanding translunar coast photograph extends from the Mediterranean Sea area to the Antarctica south polar ice cap. *(NASA)*

Below: The Hubble Space Telescope's mission is to survey the heavens, studying everything within its reach—a reach that is collecting images 14 billion light-years from the infant universe. *(NASA)*

mosphere, life-giving for only three miles above the surface of a world nearly 8,000 miles in diameter. In this planet's past, beneath this onionskin layer, vast seas abounded with the elements and nutrients necessary for primitive life-forms to emerge from flora. Once they began their journey through the seas to sandy and rocky beaches, absorbing atmospheric gases and basking in sunlight that stirred the broth of chemicals and procreation, the grand march began: eating, reproducing, and dying.

On this planet there would emerge a creature with a brain that enlarged and deepened, that took advantage of bipedal form and an opposed digit. In a silent explosion that made no more sound than two snowflakes colliding in an Arctic night, the miracle happened. The brain perceived itself and in that moment the animal functioned with a mind.

The planet is Earth. The small and frightened creature in a world of powerful and hungry beasts persevered. Millions of years later, now the wielder and master of great forces, from electricity to grasping the atom, the animal was full-grown to man.

Millions of years ago, when our ancestors looked into the night sky, the sparkling, glowing, changing orbs of the heavens must have prompted an astonishing gamut of emotions. Awe and wonder led to the thirst to know more, the desire to comprehend, which in turn led these people to build marvelous machines.

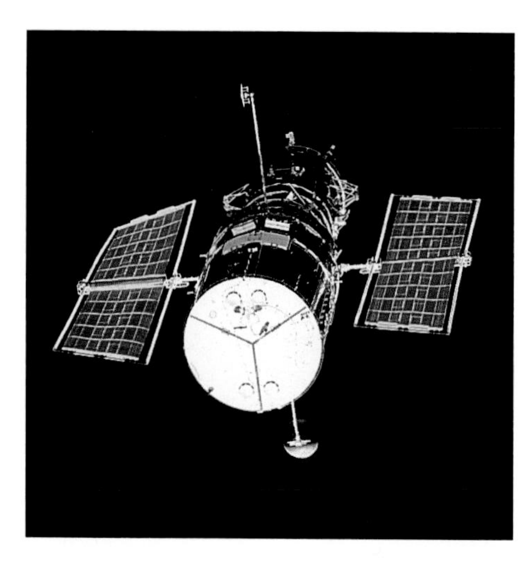

One such is the Hubble Space Telescope. Hubble, and other instruments of stellar exploration designed to detect light and en-

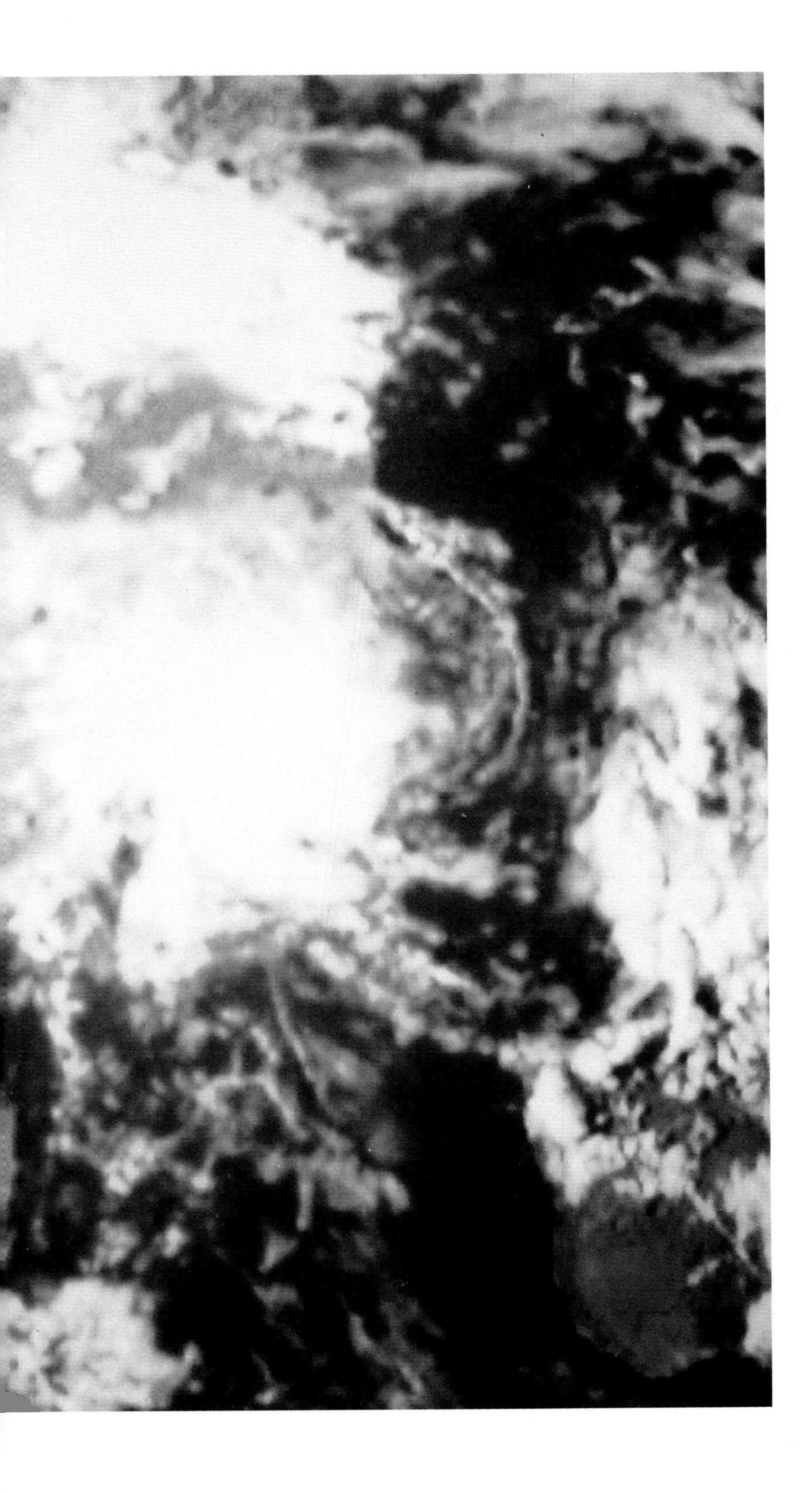

Left: This view of Madagascar, taken by the Clementine spacecraft, shows the clouds, water, and land needed to support life as we know it—the type of planet the Hubble Space Telescope is seeking. *(Naval Research Laboratory)*

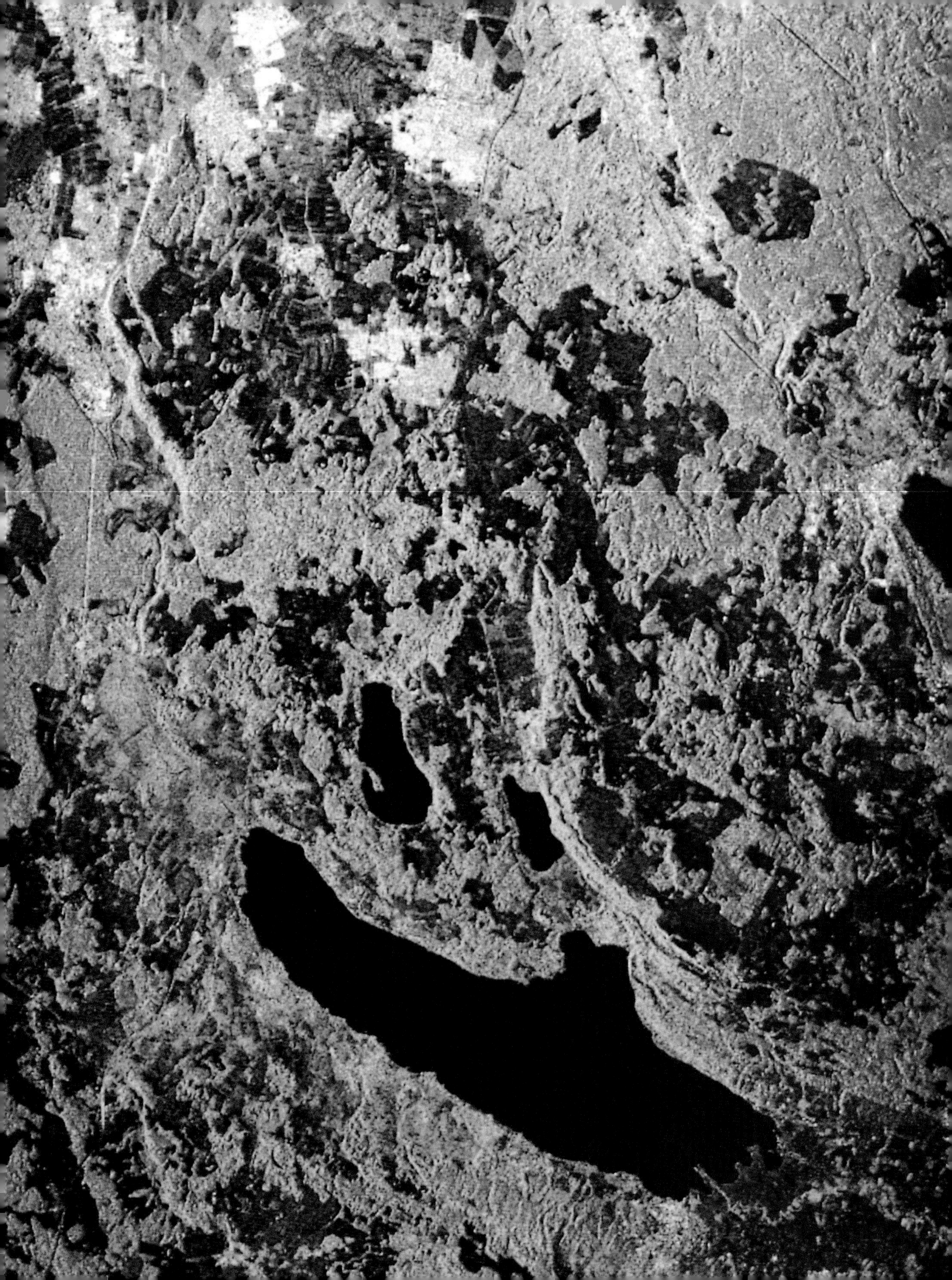

Left: It won't be easy for Hubble's optics and instruments to spot life on other planets. Even from the space shuttle Endeavour's orbit, just 150 miles in space, it is difficult to detect signs of living creatures on Earth below. This false-color composite image is of the Oberpfaffenhofen, Germany, area. *(NASA)*

ergies we see as well as those that are invisible to our eyes, have stressed one lesson above all others about alien worlds: Life will be found on other "water worlds"—planets with the same biologically favoring characteristics as Earth. Of this there is widespread acceptance by scientists. How such worlds eventually develop is open to question and speculation. The issue here is *life*, and that life evolves into intelligent, creative races.

It *will* happen. Someday we will encounter one another.

Our road map of the universe begins here on Earth because our globe is the best possible yardstick by which to measure other planets. The first key is to search for a water world. The second is that size will often prove to be critical. Compared to giants like Jupiter, Saturn, Uranus, and Neptune, Earth is diminutive—at just under 8,000 miles in diameter, Earth is less than one-tenth the size of Jupiter.

Bigger is anything but better. No form of life that does well on Earth could survive the hostile Jovian atmosphere. Stupendous gravity, bitter cold, blazing heat, violent storms, poisonous gases, and lethal radiation are just the start. There's no place to stand; a man seeking a surface would sink relentlessly through ever-denser hydrogen and helium, and finally would be crushed by gases that become immensely dense liquids.

What separates Earth from all other known planets is a marvelous cornucopia of atmosphere, temperature, water, fertile soil, a superb weather system, protective layers in our atmosphere against incoming dangerous radiation, and gravity to which we adapted so splendidly.

Right now we believe we're the only act on the vast stellar stage with advanced life-forms—which is why the Hubble team will focus future searches for equivalent planets in terms of parent star and worlds approximating the physical characteristics of Earth.

The geographical distribution, variety, and sheer numbers of life-forms on our world are so vast that our sciences and studies have never been able to catalog fully what exists about us. Every form of life is cut from the same original genetic pool that began billions of years ago, beneath solar radiation, organic molecules, and lightning mixing the primordial soup to begin a variety of life-forms.

Such vast numbers and variety have proven the key to survival. As humans, we now modify and alter the surface and nature of our world. Much of that alteration is dangerous and demands quick retraction, especially of toxic and chemical wastes poured into air, soil, and water.

Below: These four images show varying perspectives of Isla Isabela, one in the chain of Galápagos Islands, located in the eastern Pacific 750 miles off the western coast of Ecuador.

The Galápagos Islands have six active volcanoes—similar to the volcanoes found in the Hawaiian Islands—and reflect the volcanic processes that occur where the ocean floor is created. *(NASA)*

As splendid as our planet is, its ecosystem remains nevertheless fragile in so many respects that our continued existence at times seems to hang by perilously thin thread. By the time humans built their first cities about 7,000 years ago, perhaps 95 percent of every type and class of living species that ever existed on Earth was gone forever. Impacts of huge meteors and asteroids, continents splitting apart, and severe ice ages wiped out most life on Earth.

Catastrophic ages will always be a Damoclean sword over inhabited worlds. But always we return to the greater possibility of life appearing on a water world than on any other planet. Never to be ignored in appraising Earth or any other globe is the immense fortune of being in the right place at the right time.

Out of the nine known planets and several moons in our solar system that might harbor some form of life, methane-sucking though it might be, how did we gain oceans of water while the rest of the planets ended up too hot, too cold, too large, too small, too dry?

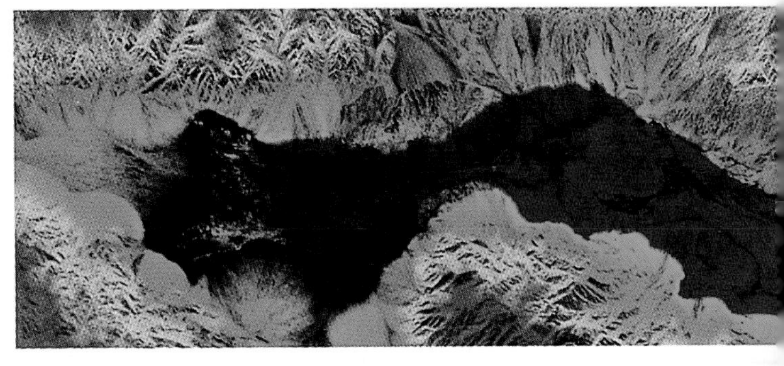

Above: Earth is at the perfect distance from the sun for living creatures to evolve—even in California's Death Valley. The SIR C/X Sar radar took these images from the cargo bay of the space shuttle Endeavour as it orbited over Death Valley April 9, 1994. *(NASA)*

Whatever the cause—accident or design—we're at just the perfect distance from the sun for living creatures to evolve. *That's* a yardstick in the Hubble manual for studying other words. To welcome life, a planet needs not only proper distance from its star, but adequately fast rotation, sufficient water in liquid from, an ongoing weather engine, reasonable temperatures, and protection from nuclear and cosmic radiation.

The recipe for the perfect planet applies to Earth. Did other worlds circling the same star ever have water? Mercury and Venus are rejected immediately, as we have seen. But the fourth planet from the sun, Mars, *did* have water, immense amounts that slashed huge channels in the surface, cut deep valleys, created rich clouds. Mars, half the diameter of Earth and with much lighter surface gravity, *almost* made it. Lacking sufficient mass to retain the lighter gases, just far enough from the sun to be much colder than

Right: Mars once had water. Water slashed huge channels in its surface, cut deep valleys, created rich clouds, and the planet *almost* made it. This true-color picture of Mars was processed by the U.S. Geological Survey from images sent back by the Viking Martian spaceships. *(U.S. Geological Survey)*

warmer, Mars clung precariously to the lip of the evolutionary cliff. Slowly the atmosphere thinned. Oxygen never poured into the atmosphere, and finally the pressure of this luckless world became so low, the temperature so cold, that liquid water could no longer exist. What water remained was locked up in frozen polar caps and underground chasms. Here is a planet that died aborning.

The comparison accentuates the great fortune of our world against the hapless short shrift of neighbor planets. So we, of Earth, become the yardstick by which to measure life-forms yet to be discovered.

About the time the last Apollo mission to the moon ended at the close of 1972, scientists and engineers looked forward eagerly to manned flights to Mars that might discover at last fossils of life-forms that struggled to evolve but died when the means to sustain that life was exhausted. Lysle Wood of the Boeing Company's Aerospace Division shared a roundtable with selected science writers to emphasize how easily two links missing in the chain of ingredients for planetary life could doom evolution to a dead end.

"If I were a man on Mars," Wood told his rapt listeners, "I might want some substances which were not native to my planet. For example, I might want a gas which would be clear, odorless, nontoxic, and would conform to the laws of a perfect gas. It would retain its chemical composition and remain fluid at all temperatures. Furthermore, it would be relatively inert and easily handled. Or I might want an incompressible, stable liquid which could convert to solid or gas. Furthermore, it would be nontoxic in all states, and storable.

"What I might want, but might never know existed, are air and water."

So simple, so elegant. So commonplace on our planet with most of its surface covered with water—oceans down to seven miles deep. Yet, when life-forms first appeared, Earth in many respects was a world we would never recognize. No familiar continents, no large animals, no free oxygen in a hostile atmosphere, no ozone layer protecting budding, fragile life. Yet it was a globe of rich potential.

Imagine a future when we travel to alien planets. One such world captures our interest because it is the same size and mass of Earth. This yet-to-be-discovered globe is wracked with volcanic eruptions and devastating earthquakes. Poisonous gases swirl in deadly amounts. Geysers and fierce storms add to the deadly violence. There is no life-giving oxygen to be found. Meteoroids and asteroids rain down from space in nonstop bombardment, split-

ting the land and unleashing massive tidal waves. A huge moon hangs perilously close to the planet, raising engulfing tides. It is a planet of monstrous devastation, so hellish it seems life could never begin here—yet that was Earth so long ago!

It is from all the ingredients mixing on a much younger Earth, lethal and destructive, that life evolved in the protective embrace of oceans and then began the perilous trek to dry land. Science gives us a clue to what we may encounter as alien life—a clue dredged from our own past. Proteins and their amino acids are not only universal components of all living things, but also are the most abundant organic chemical components of living things. Except in plants, about half of the dry weight of living matter consists of these molecules.

The amino acids of human bodies match those found in dinosaur bones and in fishplate hundreds of millions of years old. Many of the 18 amino acids that persist in living things can survive in their original form for hundreds of millions of years, far beyond the time period needed for even primordial slime to produce and reproduce.

What, then, is the yardstick by which we judge life itself on our world and the alien globes to be visited?

Life is a system capable of storing information, and then using that information to duplicate itself.

Then how do we determine if life-forms exist on a planet without "getting down among them," as our lunar astronauts said of final approaches to landing on the moon? Any similarity to ter-

Left: Clearly visible in this picture are the steep slopes on the upper flanks of the Mount Pinatubo volcano in the Philippines. The volcano's main crater was produced by the June 1991 eruption. The red color on the high slopes indicates ash deposited during the eruption; the dark drainages represent mud flows that flood the river valleys following heavy rainfall.

The picture was taken by the SIR C/X Sar radar on board the space shuttle Endeavour. *(NASA)*

Below: This color picture from the Galileo spacecraft shows the limb of Earth looking north past Antarctica. *(NASA)*

restrial yardsticks can be sorely lacking. We might not recognize alien life-forms—*nor they us.*

Astrobiologists take glee in photographs from the Galileo spacecraft when it passed over Earth at a distance of only 600 miles. Scientists poring over the photos were astonished to learn that in not a single picture was there the first evidence of intelligent life on this world.

Biologists estimate that if we weigh every living creature on Earth, then 15 to 20 percent of that total weight would be made up of ants. After sharing this planet with ants for millions of years, we have still not identified all the species—and that's with groveling within the dirt to find the creatures believed to number in the *trillions*. So an entire planet filled with trillions of animals might not show, from space, the sketchiest sign of life.

The sheer adaptability of life across our home world bodes well for finding life elsewhere. If our planet is any indication, there's an incredible variety of life on other water worlds.

Spiders casting parachute sails of silk soar 60,000 feet high, riding thunderstorm updrafts like balloonists. At the opposite end of that scale, shrimp and flounder thrive at the bottom of the Challenger Deep in the Pacific Ocean's Marianas Trench. We live under air pressure of 14.7 pounds per square inch at sea level. Sea creatures found at ocean's bottom live, feed, and procreate under pressure of *16,000 pounds* per square inch.

Animals survive under desert heat of 155°F, while other organisms endure still-air cold of minus 100°F. Surface extremes mean little to the creatures living within 140 million square miles of ocean. They have 329 million cubic miles of water to feed and protect them from the dangers of dry land.

But the survivability of land life-forms promises all the more hopes for life on other worlds. Trees live through winters with cold lower than minus 100°F. and thrive with the change of seasons. Great redwood trees live for thousands of years. Some vegetation procreates without ever being touched by water. Plants with aerobic respiratory systems seal themselves off from the environment and live solely on sunlight, using photosynthesis to produce needed liquid and oxygen.

And in bubbling hot springs of deadly liquids, algae and lichens live for years in temperatures of 400°F. Scientists note that cockroaches can withstand a million times the nuclear radiation that kills humans—and go right on doing what roaches have done for hundreds of millions of years.

What nature can do, science can mimic. Laboratory scientists have produced such plants as reindeer moss that safely endure

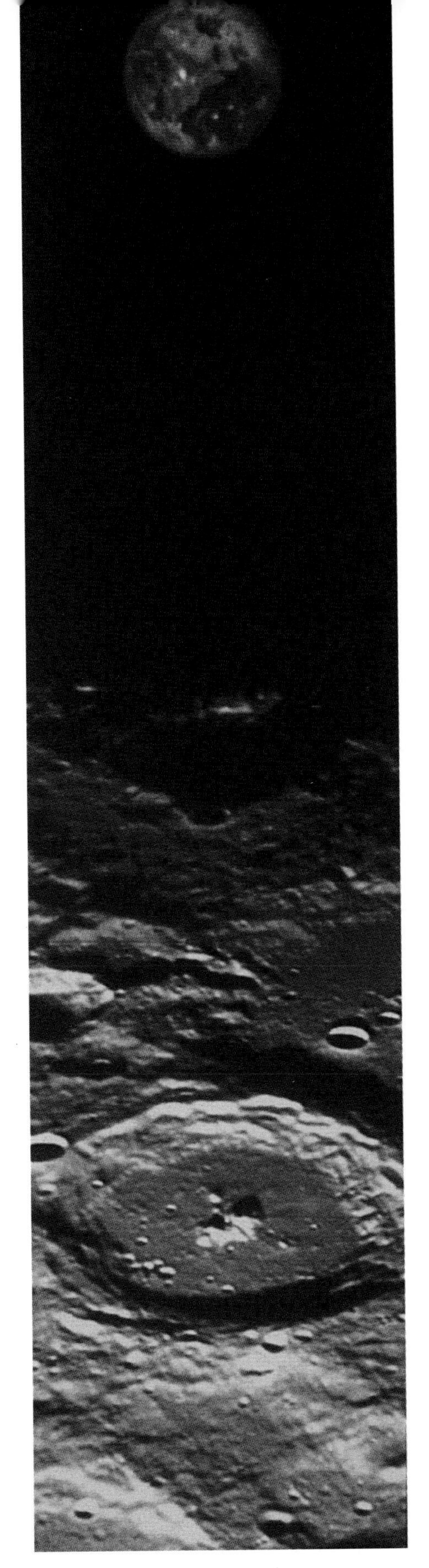

Left: The spacecraft Clementine took this colorized image of the full Earth over the moon's north pole. *(Naval Research Laboratory photo)*

4,000 times the ultraviolet radiation that in hours would char human skin to third-degree burns. A moss has been bioengineered to endure a thousand times the gamma radiation that kills humans—and remain unaffected.

There is yet another yardstick with which to measure distant planets. Earth undergoes intense radiation bombardment from the sun, which sends floods of lethal particles toward us. Against this constant blast wave is an invisible barrier—a powerful magnetic field. Earth's core is molten; we are the densest world in the solar system. Density, and the materials that make up this planet, generate this magnetic field that protect us.

Should the sun's highly penetrating energy lash Earth without the protective magnetic field, life would be eradicated, except for creatures burrowed deep within rocks and far down in ocean depths. Humankind would perish.

If we could see the clash of forces about our world, we would take solace in a magnificent glowing arc of blue-white light 50,000 miles about our planet. Here the solar cyclone first smashes into our magnetic field; most is deflected. But some of the barrage, still in lethal force, blasts another 10,000 miles closer, crashing against the magnetopause 40,000 miles above our atmosphere. It is a ceaseless battle of stellar bullets and blast waves against the magnetic shield that diverts away most of the killing rays.

But not all. The solar bombardment that our eyes cannot see, instruments detect as gossamer flame and needle streamers of energy pulses. A fraction of the bombardment roars against Earth,

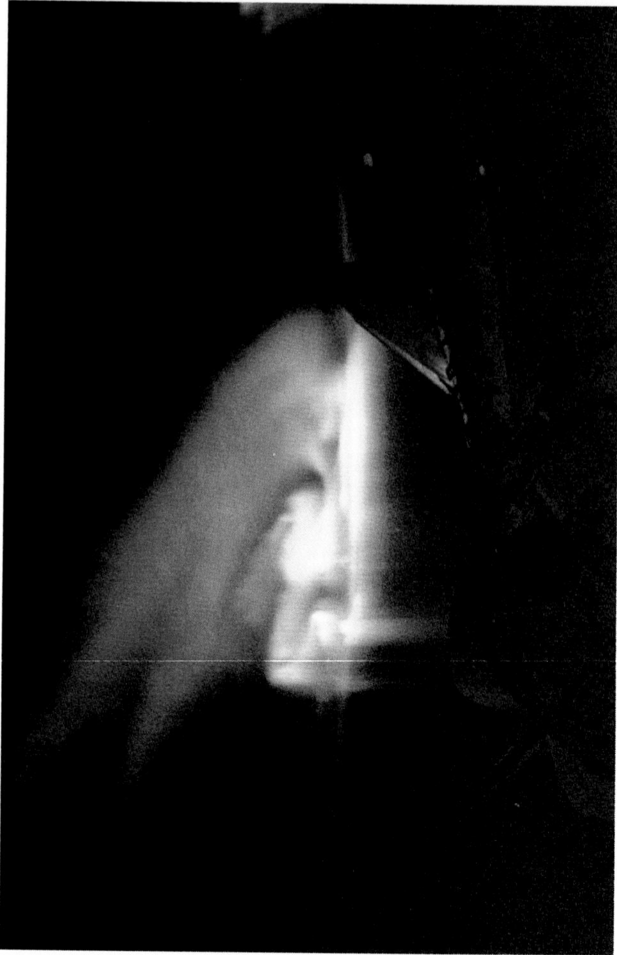

Above, left: An active auroral display is viewed across the cargo bay of the space shuttle Atlantis. The multiple auroral arcs are indicative of magnetic sunstorms caused by the intervention of solar wind with Earth's magnetosphere. The red and green colors are caused by emissions from different atmosphere gases.

This photograph was taken by the shuttle Atlantis' crew on mission STS-45 in late March 1992 over the south Atlantic Ocean. *(NASA)*

Above, right: This view of the Aurora Australis shows a band of airglow above the limb of Earth. The Southern Lights are most intense in the spring and fall of the year. This picture was taken by shuttle Discovery's STS-39 crew. *(NASA)*

but the magnetic field bends most of what remains in great curving sweeps around and away from us.

There are other intruders aiming destruction at Earth. For billions of years our world has been the hapless target for ice, dust, iron, and rocks of all sizes smashing into the atmosphere. Most of the cosmic bombardment is incinerated within the onionskin layer of air. Cosmic barrage is our planetary way of life, day and night, visible sometimes even in daylight but more often seen as blazing streaks through our night skies.

Sometimes a meteor explodes from heat and pressure, transformed into a bolide of dazzling green. If the meteor ends its existence in a fiery blast above earth, penetrating atmosphere in a shallow death dive, we hear deep booming thunder for hundreds of miles in every direction.

There exists an eerie *oneness* between the upper reaches of Earth and along the surface. From the polar regions the great Northern Lights, the Aurora Borealis, the Southern Lights, and the Aurora Australis flash high over the world. They are the sun setting aflame the electrical parts of our atmosphere as if a huge neon coruscation were switched to ON. Bands of light become streamers, banners erupt into swirling eddies, waves of light snap out and back like ghosts on a cosmic stage. And it all takes place in silence.

Far below, on some nights, darkness wraps the world while ground temperature is still warm. With gloom comes a plunge of temperature. Moisture in the air and on the ground freezes with astonishing speed, so swiftly *it can be heard*, faintly, wonderfully, in crisp tones of tiny ice bells fracturing and expanding.

It is Earth music.

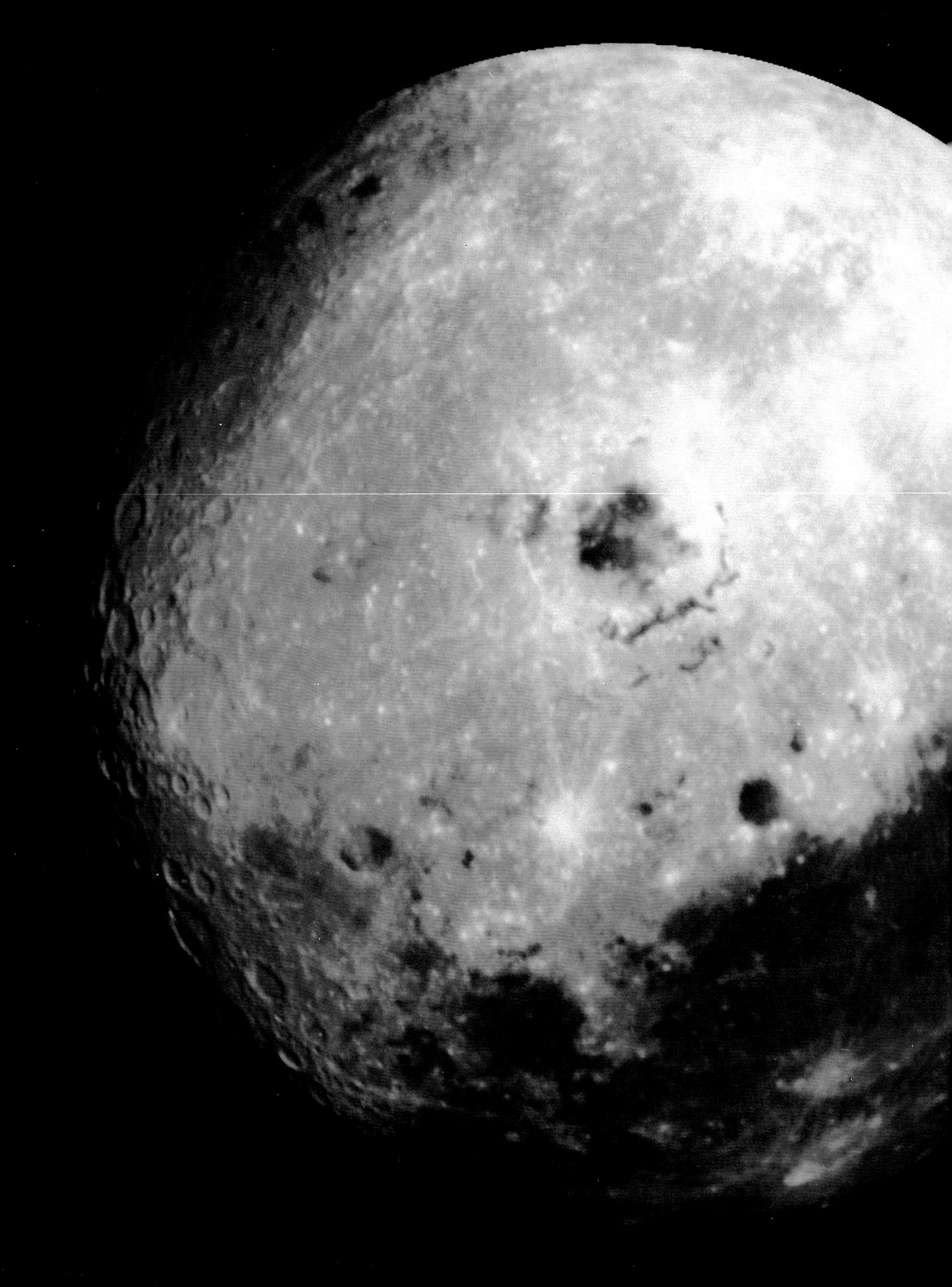

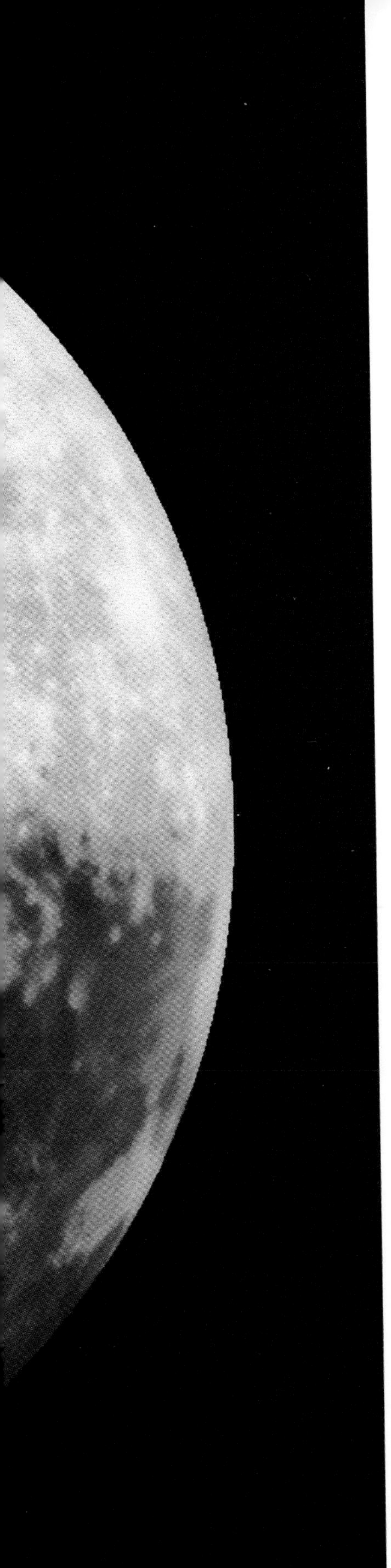

5 HARVEST MOON

It is a bone-dry world stripped of even the faintest trace of water, lifeless, barren, silent, and an unwitting target for ceaseless bombardment by pebbles, small mountains, and comet debris. It is washed in fierce radiations from the sun and cosmic barrage from distant stars. A small world, its diameter of 2,163 miles is less than the distance from Los Angeles to New York.

This is our moon, hammered into a desert of craters and jumbled heaps of boulders. Billions of years of punishment from sun and stars have pounded its terrain to soft curves and swells. Dust covers great sheets of ancient lava. Dark and foreboding, crusted, split with gorges, rills, and deep canyons, it rolls silently about its host planet a quarter million miles distant without the slightest hint of life. Not even the smallest of insects scurry across its dusty surface.

But it wouldn't stay that way.

In the summer of 1969, life came to the moon when a human being climbed down the ladder of a spidery craft covered with wrinkled gold foil. Long metallic legs and rounded foot pads splayed out from the chunky machine. Through the end of 1972, eleven more men would follow, bringing with them shining instruments to measure, sift, electronically sniff, and collect the secrets of this desiccated globe.

For the first time ever, men walked and rode across the surface. Their footsteps, the wash of rocket flame for descent and departure, the soft bouncing of small two-man hot rods to traverse the rock-strewn dusty surface, were the first impacts and sounds of human beings. They came to the moon to study and learn, to wander through this time machine of an ancient world. To aid their studies of what the moon might be like far beneath its surface, to take its temperature and peel away its secrets, they sent instrumented probes to smash into dust and rock. After men left robot science stations on the lunar landscape, they aimed huge rocket stages to crash into the luckless globe. The crust of the moon shivered with the blows, shock waves raced far beneath the craters and frozen lava, and the moon began to yield the secrets it had locked away for billions of years.

Of all the great rewards of the long ascent from one world to another, the most unexpected took place when men lifted their

Previous page: This color image of the moon was taken by the Galileo spacecraft at a range of about 350,000 miles. *(NASA)*

Right: America's first astronaut in space, Rear Admiral Alan Shepard, flew this moon lander he named Antares to a dark, lunar landscape called the Fra Mauro highlands. *(NASA)*

Below: Moondust covers great sheets of ancient lava as the lunar world rolls silently around its host planet a quarter of a million miles distant.

This view of the rising Earth greeted the Apollo 8 astronauts as they came from behind the moon on man's first orbit around the lunar surface. *(NASA)*

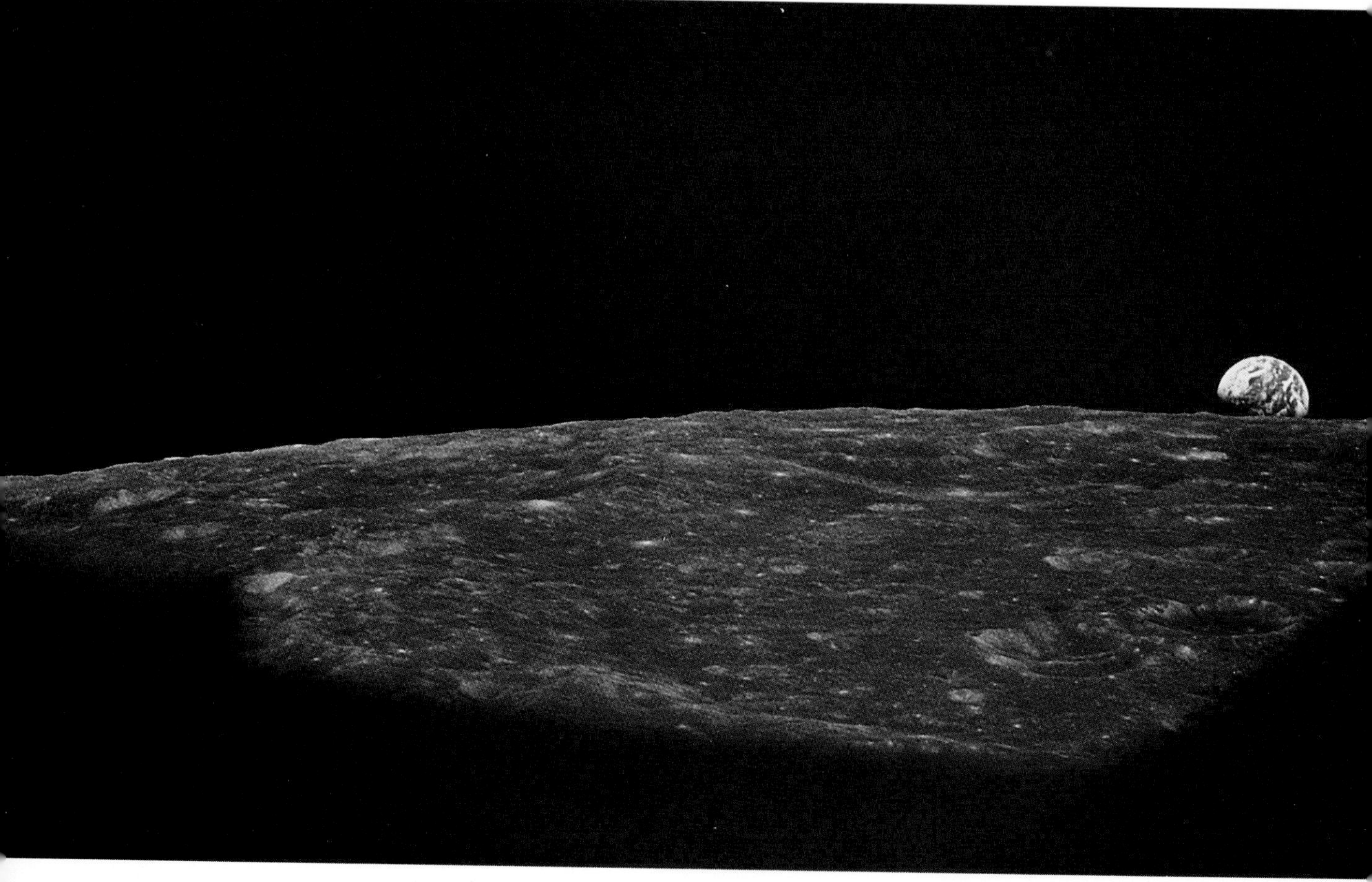

UNITED
STATES

Left: Major General Charlie Duke, lunar module pilot of the Apollo 16 mission, collects lunar samples here beside his moon buggy near Stone Mountain at the Descartes landing site. *(NASA)*

Above: Our home world, the spaceship Earth. *(NASA)*

eyes from the crater walls and sloping hills to look upon a wonder they had never conceived.

Suspended in space and time, a shining blue and white marble greeted their eyes, the only color they could see no matter where they gazed out into the cosmos. Their home world, the spaceship of all humanity called Earth.

These moments in time and space were the harbinger of all man's voyages beyond his home planet to come. This was the first brief period during which the human foot left its imprint on a world other than Earth.

In this epochal slice of time, what would be remembered by the participants and the awed observers on the home planet receiving live television pictures—with a delay of one and a quarter seconds for the video and audio signals to travel at the speed of light back to Earth and be retransmitted to home television sets—was *not* the sights of the moon. Imposing as these were, they developed a familiar similarity to them quickly.

What no one really could forecast was the tremendous impact of looking into bottomless, endless dark to gaze in awe at the glowing world of man. Never until this instant did the human race share so completely in a sight that would begin the rethinking of much of the world's population. Earth was beautiful; Earth was incredibly fragile, a spherical spaceship carrying every living creature

Right: Gene Cernan, seen standing here before his lunar landing module and his lunar rover after raising the American flag, says, "I don't like being referred to as the last man to walk on the moon. I like to think of myself as the most recent visitor." *(NASA)*

Far right: The Apollo 14 lunar module "Antares" can be seen photographed against a brilliant sun glare in the background. A bright trail was left in the lunar soil by Astronauts Alan Shepard and Edgar Mitchell's two-wheeled "golf cart" as they explored the dark Fra Mauro landing site.

Before leaving the moon, Alan Shepard would become the only person to hit a golf ball across the lunar landscape. He would later tell us, "There are lots of bunkers up there." *(NASA)*

spawned by millions of years of evolution.

Experienced astronaut and commander of Apollo 17 Gene Cernan had been on one earlier Apollo mission, but had not landed. Cernan, the "last man back" from the moon landings, offered his memories and feelings to the writers for these pages.

"We journeyed—that's the word for it—into and through a quarter-million miles of utterly black space to that small planet we called the moon. And I was caught by surprise; stunned, almost. Here I stood, now on dusty ground where a man became detached from himself. Everything you have ever learned and understood moves away from you. Here, for the first time, I encountered a wholly new experience. What I understood of time became the fourth dimension and where reality itself was dreamlike. . . .

"To stand on that dusty surface in sunlight, gazing at that glowing planet surrounded by the deepest blackness one might ever conceive, defies imagination. Watching the sun set on one continent and rise on another is less looking across space than peering through time."

Gene reflected deeply and talked slowly, drifting back to those incredible moments when first coming around the moon in his Apollo command ship, "being exposed to an awesome tapestry painted across the velvety black of space. You know, coming up and around the moon in lunar orbit, the sight of distant Earth is really startling. There's no way really to describe the utter black of space. It is a complete, overwhelming, utterly black upon black upon *black*. And you take the time to really *look* at the world you left not long before, against this velvet blackness, the spherical jewel, stunning blue, and whorls and tendrils of white clouds.

Above all else you sense the *sphere* . . . you had the feeling you could see around the curving edge of that small, distant, stunning world, and you knew, above all, that this was *real*."

Cernan's was but one of six successful manned lunar expeditions to carry out scientific studies. From 1969 through the close of 1972, the twelve men on these expeditions represented the entire human race on an alien world.

Above: Geologist-astronaut Harrison Schmitt heads for a selected rock to be brought back to Earth from the Apollo 17 Taurus-Littrow landing site. *(NASA)*

There was a widespread survey across different areas of the moon. Inherent in these geological studies, collecting rocks, taking thousands of pictures, setting off explosive charges for seismic studies, was the search for the origin of the moon. The hidden secrets for how and when the moon appeared so long ago would be found in the 842 pounds of rock samples twelve lunar visitors returned to Earth.

In these studies would be discovered a vital lesson for the scientific teams that would soon aim their most sensitive telescopes and instruments—especially the managers of the Hubble Space Telescope—at distant planets and moons. That lesson is clear: We will never know the true nature of an alien world until we "get down among them" and walk the surface, sample, dissect, collect everything by human hand.

The moon floats in the sky as the lodestone of planetary evolution in our solar system. Less than a quarter million miles away, it is the perfect other world to study. It is close, it has no cloying atmosphere, and powerful telescopes look into craters and down canyons and along mountain ridges. There should have been little question about another world right at our doorstep.

It didn't work that way. Rather than being accurately informed of the moon, its composition and its history, the scientific data collected about Luna was a hodgepodge of guesses, inferences, and data that were just plain wrong.

The arguments across the scientific meeting table were bitter and contentious. Gross figures were disputed on every side. We had theories by the truckload, and precious little data on which we could rely. No one could say for certain if the moon had active volcanoes—or if it *ever* was active volcanically. Was there a magnetic field? Did it have a molten or a solid core? What gouged the craters in the surface? One scientific group insisted the craters were volcanic, another group championed meteoroid strikes, a third group said it was a bit of everything.

Prior to the first manned landing, Professor Thomas Gold of Harvard insisted the moon was a death trap for unwary astronauts because it was covered with dust more than a half mile deep. Fortunately, the Apollo landers sank only an inch or two before resting on firm ground.

Dr. Fred L. Whipple of the Smithsonian Astrophysical Laboratory predicted, "Loose dust on the lunar surface is practically nonexistent." He said that under booted feet the "surface should be 'crunchy' and allow minimal imprint."

The moon walkers' boot prints etched clearly in the lunar soil, and the dust that subsequently covered pressure suits, to the great

Below: Astronaut Alan Shepard raises the American flag, with his dust-covered boots etching only shallow footprints on the moon. *(NASA)*

distraction of the astronauts, put that theory to bed.

Astronomer V. A. Firsoff, who spent his entire adult scientific life in lunar observation, claimed the moon was "rough ground, largely impassable to any landborne transport, and it may well prove still more impassable than it looks." His adherents promised a lunar surface of jagged rocks that would slash astronaut pressure suits to ribbons.

Right: In this photograph Apollo 12's Lunar Lander can be seen in the background while Astronaut Conrad examines Surveyor 3's television camera before detaching it to bring it back to Earth. *(NASA)*

The sight of battery-powered lunar rovers bouncing about the moon with gleeful ease, to say nothing of astronauts who fell and jumped onto that surface without tearing up their suits, disproved those theories.

There is one lesson that stands out above all the others that must be considered when we search for life, first with Hubble and then with robot and manned spacecraft, especially the latter.

Always expect the unexpected.

Dr. Harold C. Urey of the Institute of Technology and Engineering at the University of California warned that navigation to any part of the moon was a problem we would never solve; we'd just have to land anywhere it looked safe. Urey castigated engineers' claims that we could first land robot craft and that the manned ships would follow. He insisted such navigational feats to find and land by another ship were pure hokum, that it would be "necessary to spend some years exploring the moon before it could be found."

Apollo 12 lifted off for the moon from Cape Canaveral in rain and was struck by lightning. In their typically nonchalant manner, Astronauts Pete Conrad and Alan Bean reset circuit breakers and computers and landed on the moon. Five hours later Conrad went down the ladder of his lunar lander Intrepid. "They aren't kidding when they say things get dusty, whew!" he sang out. As his booted foot struck the moon's surface, Conrad called out to his companion and millions of people listening back home, "Boy, you'll never believe it, guess what I see sitting on the side of the crater, the old Surveyor."

Surveyor 3, a robot lander that dropped to the lunar surface two years earlier, was only 600 feet away. Dr. Urey's dire predictions about stumbling about on the way down to the moon, never knowing where they would land, evaporated in clouds of moon dust Dr. Whipple predicted weren't there. Indeed, Conrad could have landed directly alongside Surveyor 3, but had aimed for his landing site 600 feet distant to prevent rocket-blasted dust from spraying over the silent robot.

There was yet another phantom to be put to rest. Dr. Urey, responding to fellow scientists' concerns about Apollo landings con-

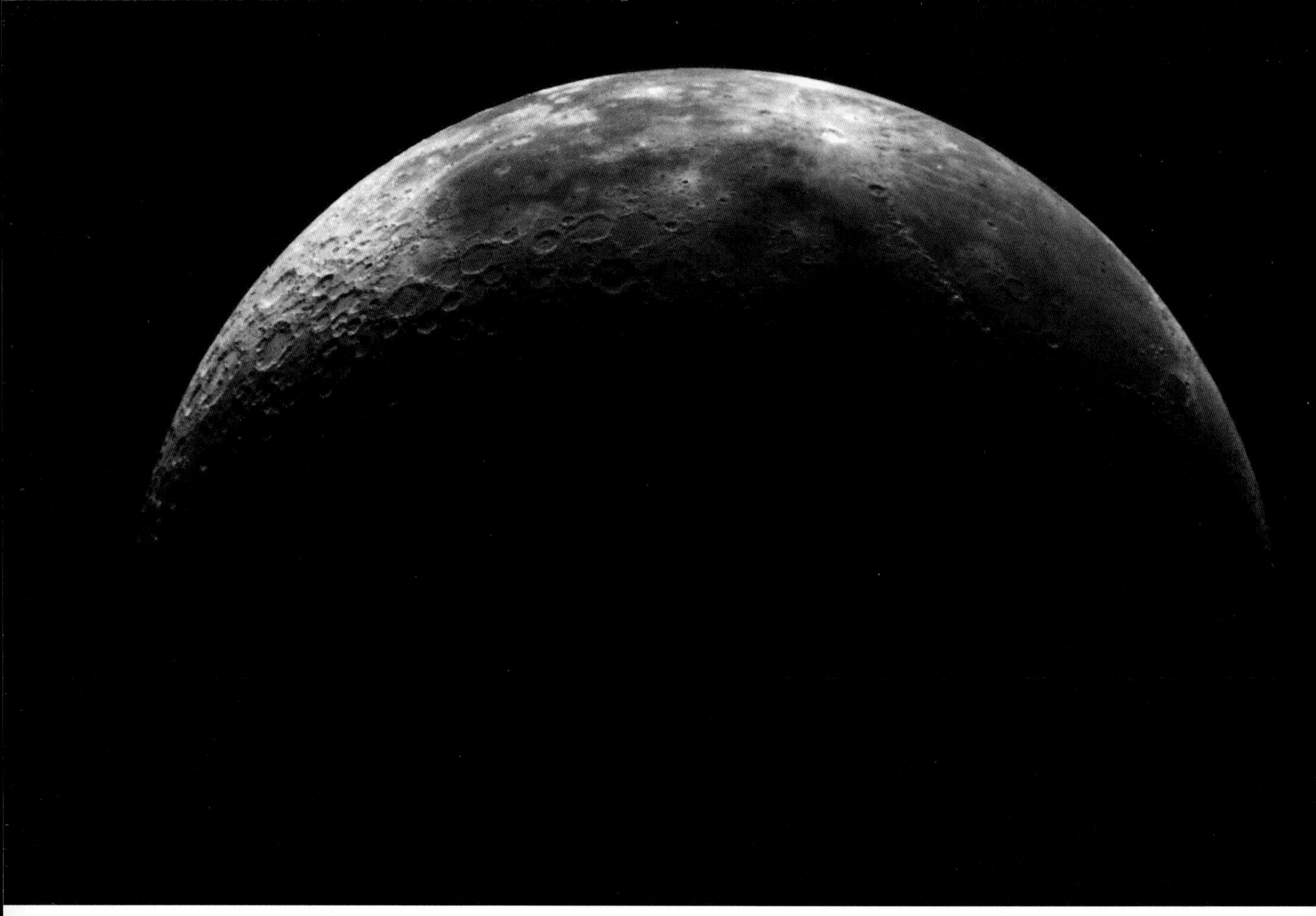

Above: This image of the crescent moon was taken by the Galileo spacecraft two decades after the Apollo astronauts had concluded their visits. *(NASA)*

taminating the moon with terrestrial organisms, stated that even were there an accident spilling earth organisms on the moon, "within one month all bacteria would be dead . . . they would dry up completely."

Surveyor 3 had been on the moon for two years in savage heat, cold, vacuum, and withering radiation. Conrad and Bean returned to Earth with several pieces of the Surveyor. When scientists studied those pieces, they were astonished to discover living organisms on the remains of the lunar lander. They came not from the moon, but had been on Surveyor 3 all through its flight from Earth and during violent rocket firings to correct its course and to land it for its long wait in lunar dust. *Streptococcus mitis*, adhering to Surveyor, lay dormant for two years; upon return to terrestrial atmosphere, pressure, and warmth, it regained life.

The six manned flights of the United States, and dozens of unmanned probes by the United States and the USSR, laid bare the concealed lunar history. From the lunar highlands came rocks between 4 billion and 4.3 billion years old. In the low areas, little or nothing remained of the original moon. A molten outer surface layer, fierce volcanic blasts, and finally a nonstop barrage by massive asteroids and meteoroids pummeled the surface until it gained a new surface of broken rocks, boulders, craters, and depressions astonishing for so small a world. At the South Pole–Aitken Basin, the Clementine lunar orbiter photographed a chunk gouged out of the moon, a chasm nearly eight miles deep—from ground level to the bottom, twice the depth of the Grand Canyon.

THE HARVEST

Were it not for the manner in which the moon was created, it is gravely doubtful that Earth would have any sign of life. The new planet, approximately one billion years old, would slowly have lost its gifts of perfect size and all the other factors that blessed this world with so rich a variety and abundance of life-forms. Earth would truly have been the sister of Venus, with incinerating temperatures, thick poisonous gases, and a naked surface of volcanoes, craters, and soil mashed beneath massive atmospheric pressure.

It began with the Earth revolving about the sun, like Mercury and Venus, without a moon. Four and a half billion years ago, from the roiling debris of solar system accretion, a huge rogue planetoid, about the size of Mars, hurtled in from the far reaches of the outer planets, accelerated steadily by solar gravity.

At enormous speed the planetoid, nearly a third the size of its target, smashed at an oblique angle into the still-newborn Earth. Life had not yet formed here; any life that might have existed would have been utterly destroyed by the planet-wrecking impact.

Earth reeled violently from the catastrophic blow. The planetoid exploded into fiery debris, after gouging a terrible, great wound in the planet. A huge mass of debris left behind a flaming, quake-wracked earth as the mixed remnants of the titanic blast whirled back into space, there to be grabbed by Earth's gravity.

Most of the heaviest elements from the planetoid, especially its iron content, remained with the now-molten Earth, beginning a long settling motion to the core of our world-to-be. As Earth cooled, it became a planet much different from before the collision. Like Mercury and Venus without their moons, Earth had rotated sluggishly. But that terrible impact sped up our planet to reach a full rotation once every 24 hours.

Again a harvest had been reaped. The flaming, vaporized planetoid had been hurled away from Earth, and settled now into orbit. In the millions of years to follow, it coalesced under its own gravitational attraction into the sphere we see today, a moon hammered relentlessly by meteoric impact.

About four billion years ago, the pummeled moon, its surface

Right: This is the only picture taken of Earth and the moon moving as celestial bodies through space. The distance between the planet and its satellite is about 220,000 miles.

This remarkable view of Earth and the moon was taken by the Galileo spacecraft 3.9 million miles from Earth, on its way to the planet Jupiter. *(NASA)*

torn open by impacts of asteroids and meteoroids, erupted in a spasm of volcanic violence that lasted well over half a billion years. When the inner violence ebbed, it left behind a moon largely covered with hardened lava. But the barrage from space continued to carve out the thousands of craters we see today.

Billions of years ago, this now-dead world gave Earth what it needed to begin life. But the moon holds yet another treasure for our future. For billions of years the sun has lashed the lunar surface with all manner of radiation. Slowly a change began in the surface material as the rain of atomic and subatomic particles showered ceaselessly into the crusted surface. Today, the moon is rich in an element extremely rare on Earth—helium 3.

Already scientific programs have begun to determine the best way to bring together deuterium (heavy water, a hydrogen isotope) with helium 3 we hope to mine on the moon and return to Earth. In great thermonuclear reactors, powerful laser beams will fuse together deuterium and Helium 3. The result will be stupendous, controlled power, far mightier than the most powerful nuclear reactors now operating and depositing on our world vast amounts of deadly radioactive waste as the price for their energy.

But with thermonuclear generators fusing Helium 3 with deuterium, efficiency will be so great that only a single ton of this lunar wonder could meet all the power needs of the United States *for a year.*

Through the gifts of a long-dead world, the lights of future Earth will burn brighter than ever for a very long time to come.

And this marvelous fuel from the moon will power the great ships of space to follow where Hubble points.

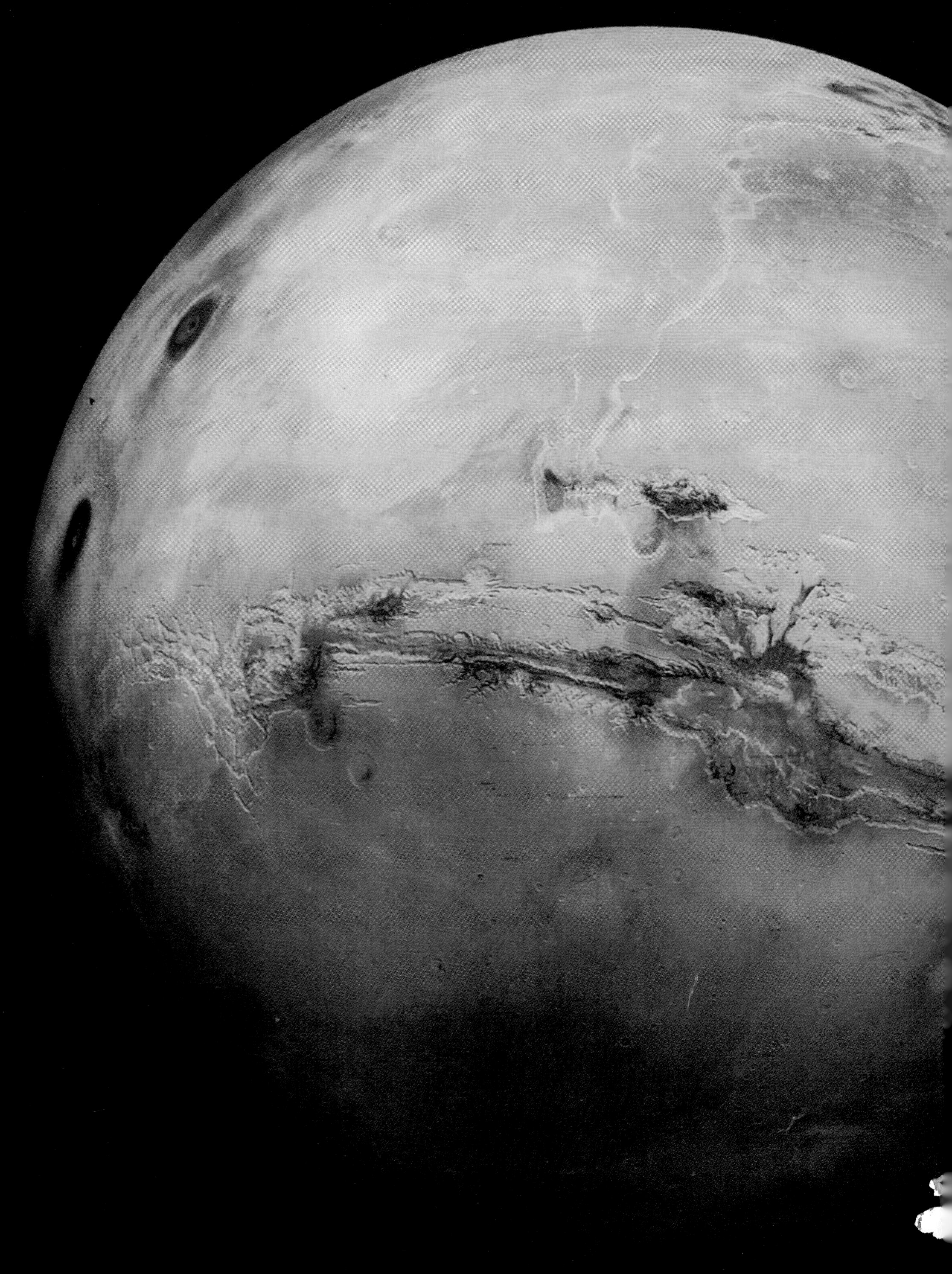

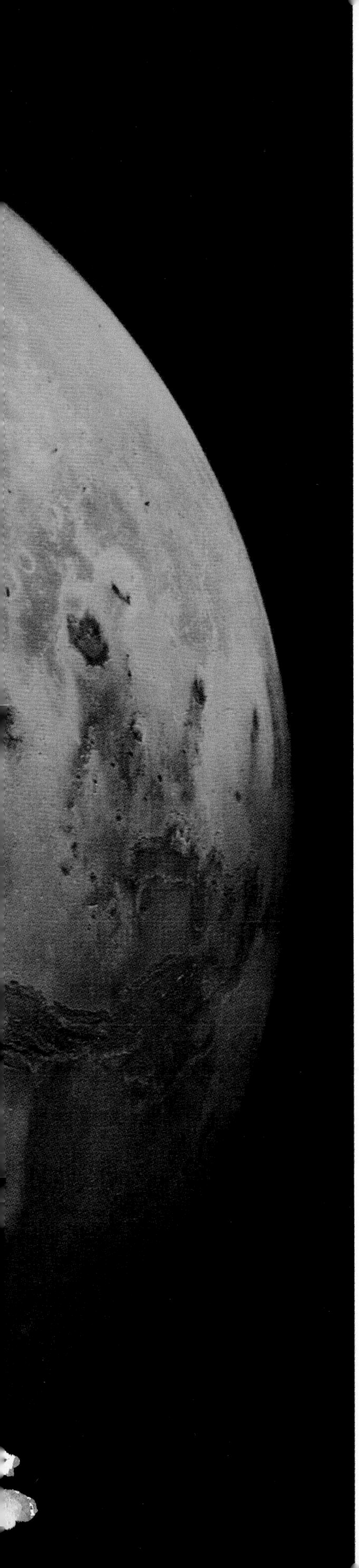

6 The Martians Are Here!

The first Martians are here, on Earth, *now*. You cannot tell them from anyone else. They live in ordinary homes, ride bicycles, play sports. They fit in within the rest of us with a casual air that so effectively keeps us from realizing who they are.

Some are less than ten years old. Others are likely in their teens. They have never been away from Earth. Neither have they lived on, or left, Mars. One day, if our plans grow to fruition, a handful of children on Earth today will be the first Martians when they land on the small orange-reddish globe less than 35 million miles from the same launchpads that sent American astronauts to the moon.

The first explorers to the moon, aliens when they landed in their skittish craft, were for the most part born in the United States in the 1920s. Those kids may have dreamed of travel to other worlds, but their hopes for realizing those dreams were bare wisps of reality. At that time, it was ridiculous. Fly to the moon? Impossible! Daft! The whole idea was insane.

But this is the 1990s, and the big difference, the all-important, all-consuming difference is that anyone who contemplates the fu-

Previous page: This Viking picture of the Martian landscape shows signs of a world that might have been. *(NASA)*

Right: This view of landslides in Mars' Ophir Chasm shows them to be more than 1,000 times greater in volume than those from the May 18, 1980, debris avalanche from the Mount St. Helens volcano. *(U.S. Geological Survey)*

ture with knowledge knows that Earth people sailing to Mars is already within our technical grasp. Commitment, funds, and spaceships are the easy part.

Decision is the powerful fuel that will take men and women from Earth to the red planet—a world that once nudged at the brink of life-forms, warm, enriched with water, filled with promise—but died.

But this is where the wonder begins. The desert world of Mars holds a promise found nowhere else in the solar system. A planet that failed can be brought to life again, nourished, fed all the ingredients to sustain plant and animal life—and become the second world for the human race to call home.

That's why Mars is beginning to fill more space on the picture-taking schedules for the Hubble telescope.

Mars is but light-minutes from Earth. It is the second-brightest planet in the night sky (after Venus), a gleaming red pinpoint in the heavens. Seeing Mars is not the problem. Resolving details of the planet seen only by reflected light demands the Hubble telescope must resolve and capture a world of gloomy features. The ice caps of Mars at times reflect light so brightly they fairly burst from pictures taken by Hubble, but that very isolated brightness pushes the rest of the planet into a blur of varying shades and tones.

The 34 robot spacecraft sent to Mars include many that were successful in collecting scientific data and taking tens of thousands of pictures. Some failed, others turned in superb performance. Of the latter, three American and two Russian ships made fully successful flights past the planet, instruments and cameras whirring and collecting furiously. Three American and two Russian spacecraft slipped neatly into Martian orbit, the first artificial satellites added to the planet's two natural moons. Two American robots made successful landings on rocky, tortured surface and performed superbly, with thousands of pictures returned to Earth before their equipment died and their power systems gave up the ghost.

The photographs taken from Martian orbit, especially from the two American Viking robots, were wildly successful. They captured Mars day and night, sunrise to sunset, covering vast deserts, volcanoes, clouds, ice fields, clumped walls, mountains, and old water-channeled riverbeds. On the surface, the Viking landers scrubbed nearby soil, with robot arms dug trenches like some child scraping dirt, turned over rocks, and took panoramic scenes of a world boulder-strewn in every direction and looking much like

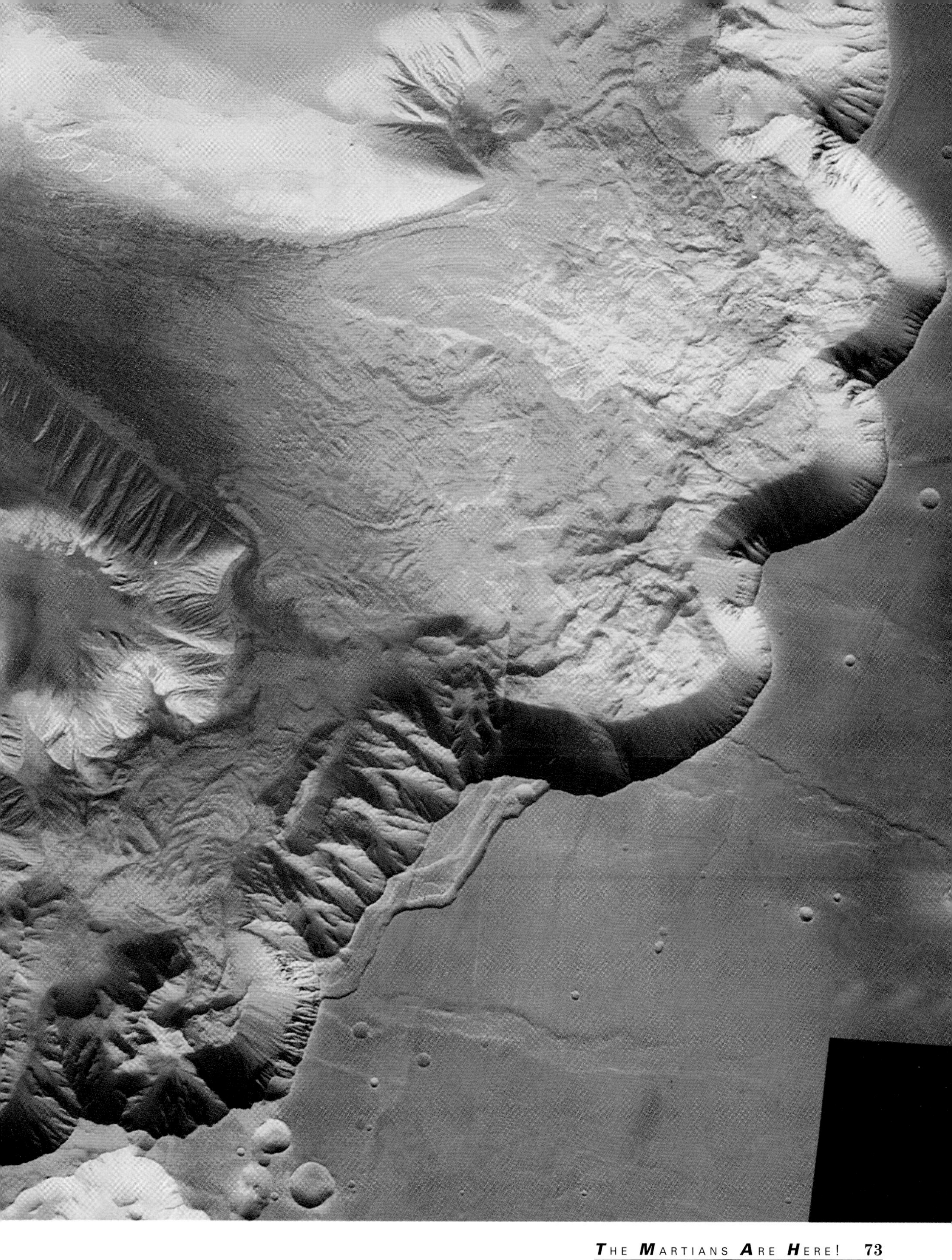

Above: The boulder-strewn field of red rocks reaches to the horizon nearly two miles from Viking 2 on Mars' Utopian Plain. Geologists believe the colors of the Martian surface and sky in this photo represent their true colors. Fine particles of red dust have settled on the Viking lander's surfaces. Color calibration charts for the cameras are mounted at three locations on the spacecraft. Note the blue starfield and red stripes of the American flag. The circular structure at the top is the high-gain antenna, pointed toward Earth to send this and thousands of other pictures.

Viking 2 landed September 3, 1976, some 4,600 miles from its twin, Viking 1, which touched down on July 20 of that year. *(U.S. Geological Survey)*

Right: The U.S. Geological Survey in Flagstaff, Arizona, developed this model of Mars from the images taken by the unmanned space probes that visited the red planet. *(NASA)*

rock-tortured fields of Earth. Most impressive of all were the scenes of Martian sunrises and sunsets with rich color gradients in color and clouds, when long ground shadows painted areas immediate to the landers with stark results.

So the question: If we accomplish so outstanding a record and plumb the depths and intimate reaches of Mars, why do we need Hubble to take pictures from so many millions of miles away?

The only way to remain current with conditions of Mars is to combine the best attributes of the robot spacecraft sent outward for detailed, closeup pictures with those of Hubble, which can sustain a sharp telescopic eye on the planet. Once robots exhaust fuel and electrical power, they're just silent, expensive junk whirling through space or abandoned on Mars' surface.

During the next several years more than 30 new robots will depart for Mars, to swing in picture-taking orbits or to land in many different areas of the desert world. The purpose of all this expensive probing is to learn as much as possible about Mars as preparation for manned flights to follow. Mars is more than a world to explore with robotic messengers; it is a planet for which plans are well under way to transform the vast frozen deserts into a new *living* frontier for men and women willing to risk settlement where life cannot be sustained without modifying—terraforming—the entire globe.

Before the robots sail, Hubble will fill the void growing ever larger since the last successful probes nearly twenty years ago, when American Vikings orbited and landed for their wildly successful missions.

But Mars is a world that changes temperature, color, seasons, and atmosphere, and the manner in which those changes take place is of critical importance for the robotic missions. Hubble can record the changes in clouds, track dust storms, and judge over long periods of time the ebb and flow of the great Martian polar ice caps.

Mars' diameter at the equator is just over 4,100 miles, slightly more than half that of Earth. That makes it tougher for Hubble—tracking and photographing a target only twice the size of our moon—but the data gathered will make it a lot easier on the first explorer to walk the alien desert. Because Mars has only 38 percent of the gravity Earth has, a spacesuited astronaut weighing 300 pounds on Earth will weigh but 114 pounds on Mars. Easy going, as our lunar astronauts found when they bounded in kangaroo hops across that dusty world.

This fourth planet out from the sun swings as close as 34.6 million miles to tracking telescopes on Earth. When both planets

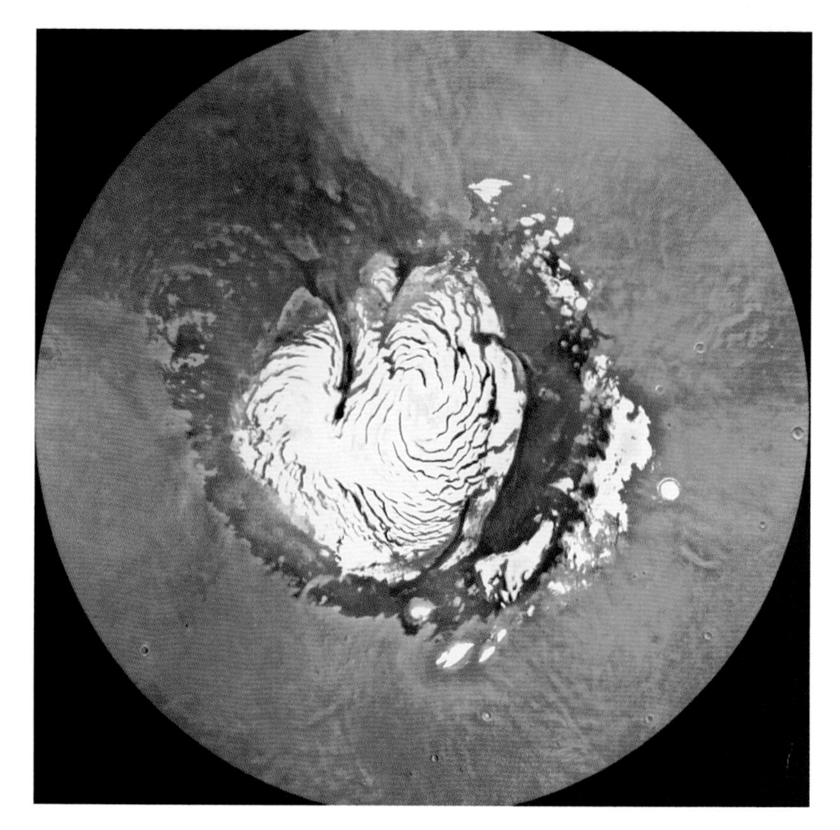

Right: The Hubble Space Telescope can record the Martian "weather" and analyze over time the shifting northern polar icecap as seen in this true-color digital mosaic of Mars. *(U.S. Geological Survey)*

are at opposite sides of the sun, picture-taking is less rewarding—the two worlds are then 238 million miles apart.

Those numbers aren't as critical as Mars' distance from our star. Mars averages 141,650,000 miles from the sun—nearly 50 million miles farther than Earth—which means it gets less than half the solar radiation in the form of heat that we receive. When you bring together these pertinent points—distance from the sun and a fraction of atmosphere—what we find is a planet starved for warmth and pressure. And though Mars has seasonal variations just like our own world, its weather can make Antarctica positively charming.

Mars has one feature denied us—two moons. Few sonnets will be written in the future about those moons compared to the avalanche of literature, poetry, and sinister attributes afforded our moon. Phobos and Deimos are lumpy, misshapen hunks of rocks resembling huge baked potatoes stabbed repeatedly with huge forks. Phobos stretches only thirteen miles from one end to the other and Deimos is even smaller, with a diameter of less than eight miles.

Decades of tracking reveal Phobos racing pell-mell about Mars only 3,700 miles above the planet—three orbits in just one Martian day of 24.6 hours. A moon that circles its parent world in a time period shorter than the rotation of that world is an absolutely unique phenomenon in our solar system.

Phobos is falling inward toward Mars, and in a mere 15 mil-

Below: Nearly 50 million miles farther from the sun than Earth, Mars is a cold, inhospitable planet. *(U.S. Geological Survey)*

lion years, barring any unusual effects upon the battered potato-like satellite, it will thunder into the Martian surface with catastrophic effect.

Predictions of astronomical affairs by world-renowned astronomers often produce great embarrassments. Russia's physicist I. S. Shklovsky judged the orbital decay of Phobos as proof it was a great artificial satellite put in orbit by long-departed Martians. "Phobos," declared Shklovsky, "is hollow inside."

Then American and Russian instrumented probes sortied to Mars to dispel this idea that an ancient race had lofted an orbital greeting card of such stupendous dimension.

The first intelligent beings on Mars will come from Earth.

An astronaut standing on the fourth planet will never want for bizarre and striking vistas. Almost everywhere is a mangled surface of jumbled rock and rippled sand dunes. In its best moments Mars is bitterly cold, an arid and lifeless desert. A wind so incredibly thin it equals Earth's atmosphere at a height of 125,000 feet still

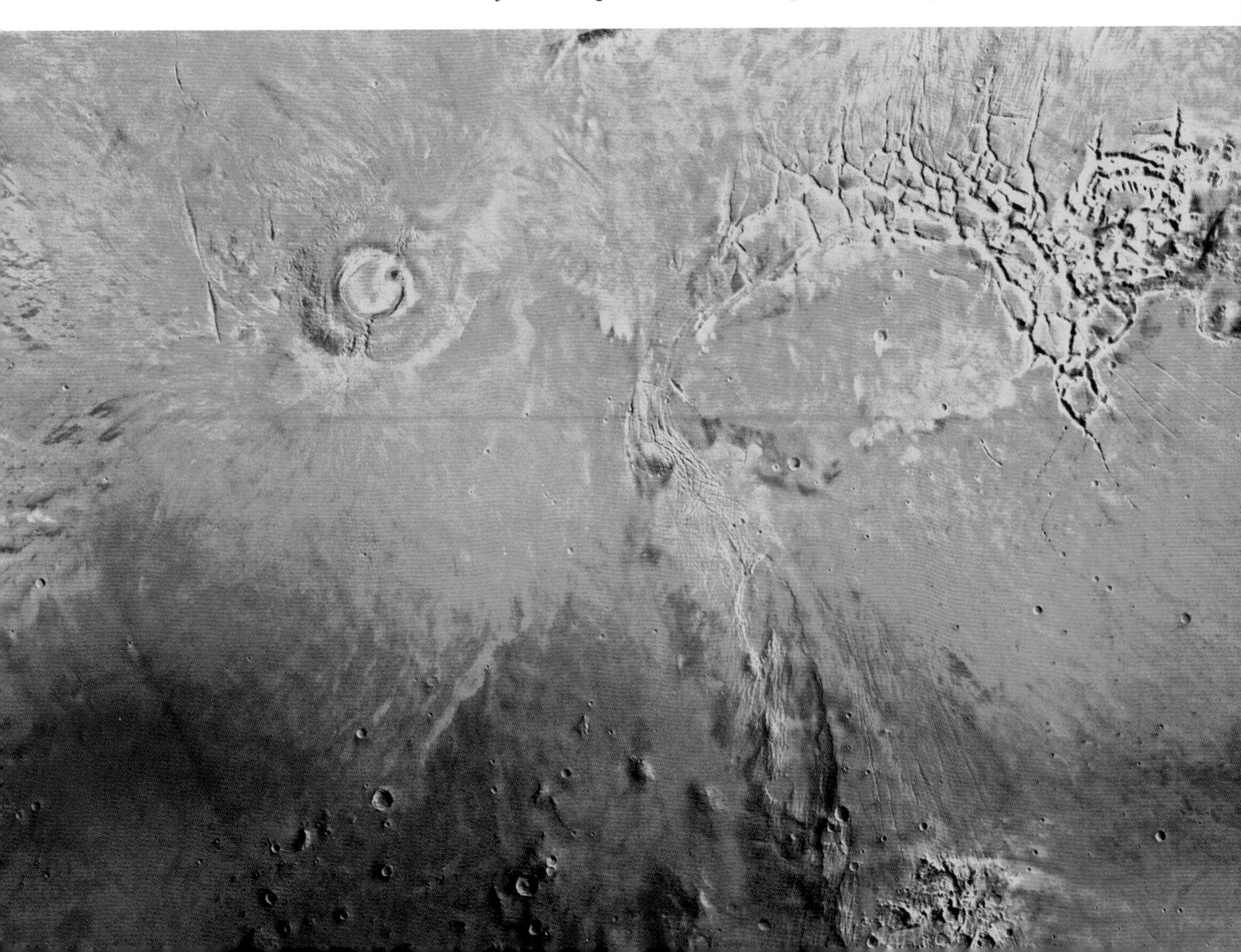

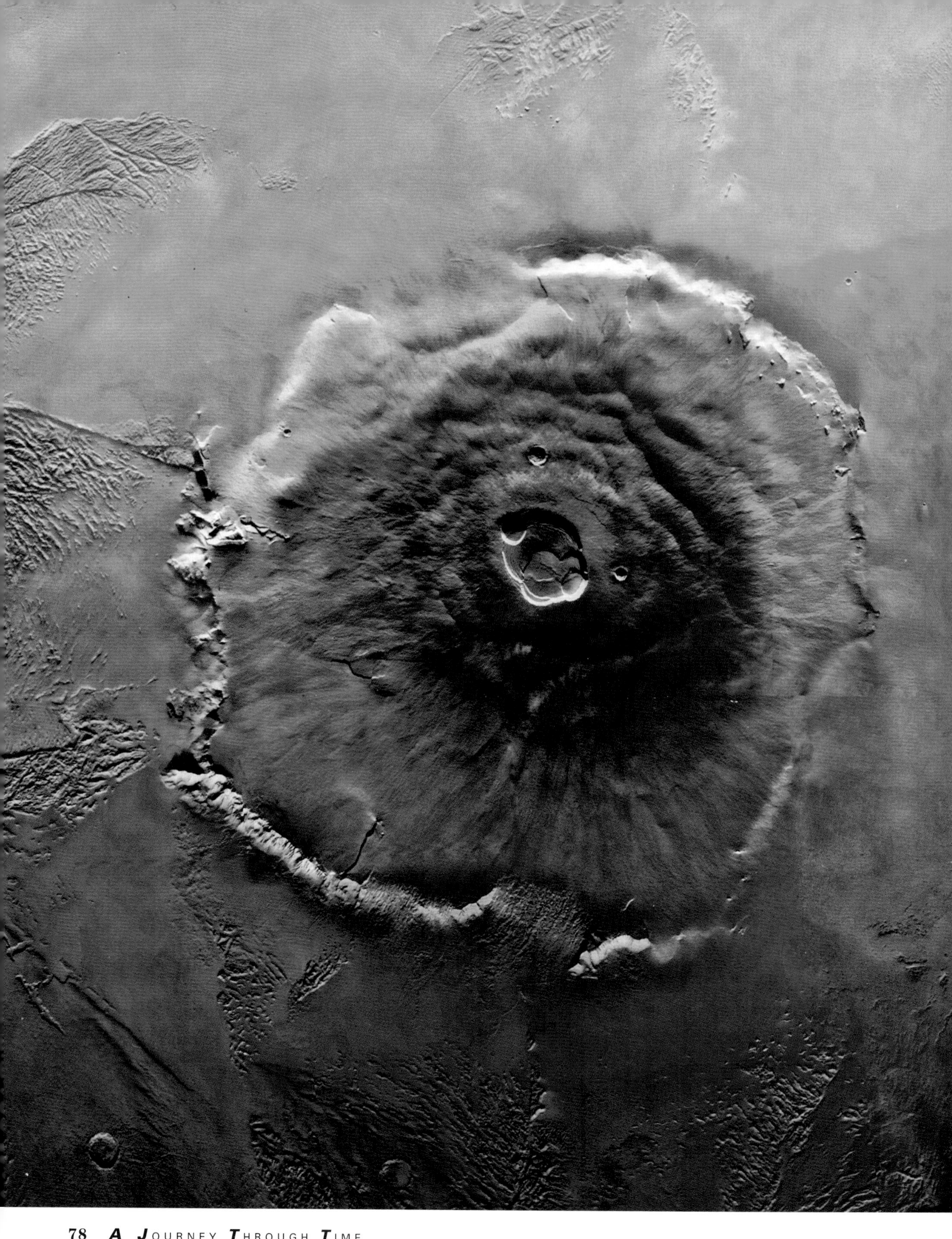

Left: Shown here is a digital mosaic of the great volcano known as Olympus Mons. *(U.S. Geological Survey)*

Below: The Martian surface seen in a photo sent back by Viking Lander 1. *(NASA)*

raises enormous dust storms so strong that sometimes we see from space a planet of only dust.

Yet through that dust our astronaut on Mars could still see the sun—two-thirds its size as the star appears on Earth—and if he flew across Mars, our astronaut would find the greatest known volcano in all the solar system, Olympus Mons.

Olympus Mons towers 90,000 feet into the Martian heavens, more than three times higher than Mt. Everest; its base sprawls over an area the size of Missouri.

Mars' bitter cold and rarefied atmosphere precludes the presence of free water, but billions of years ago the planet had rivers and great seas. Larry Helfer of the University of Rochester believes the low-lying Arcadia, Utopia, and Elysium plains of the northern hemisphere were once covered with water nearly a half-mile deep.

At that time, the younger Mars contained a thick and warm atmosphere. Erupting volcanoes hurled enormous amounts of gases across the world, heating deep permafrost that melted in flash floods to cut deep riverbeds through the land.

One billion years ago volcanic activity subsided. Low gravity could not maintain lighter gases; as they bled off into space, Mars cooled swiftly. Lakes and seas froze. In a geologically short time, the atmosphere became so thin and the cold so great that surface ice transformed into vapors spread by winds. Clouds floated about the dying world and soon released their moisture as snow. Where the snow fell on volcanoes still simmering from inner heat and warming the surface, there came a final gasp of water gushing down volcanic slopes. Today the deep gullies and valleys are frozen, silent reminders of a planet that was once warm, wet, and on the edge of life. Deep beneath the polar caps, water from long ago remains silent and frozen.

The Martian nights and winters drop surface temperature to minus 200°F. Then, at the height of summer, along the equator, the surface temperature rises barely above the freezing point.

Mars shows only pitiful remnants of a once lively planet. The carbon dioxide atmosphere with but the faintest trace of water vapor still condenses into clouds that sail high and silent above the lifeless world. Sometimes they spin and swirl about the high volcanoes, spreading ragged streamers before the winds. And in the lowlands, when the distant sun warms craters and old channels, frost appears, briefly, as if struggling to regain former glory.

One great mystery remains unsolved.

On July 25, 1976, the Viking Orbiter 1 planetary probe swung over Mars 1,162 miles above the Cydonia Plain, a barren area,

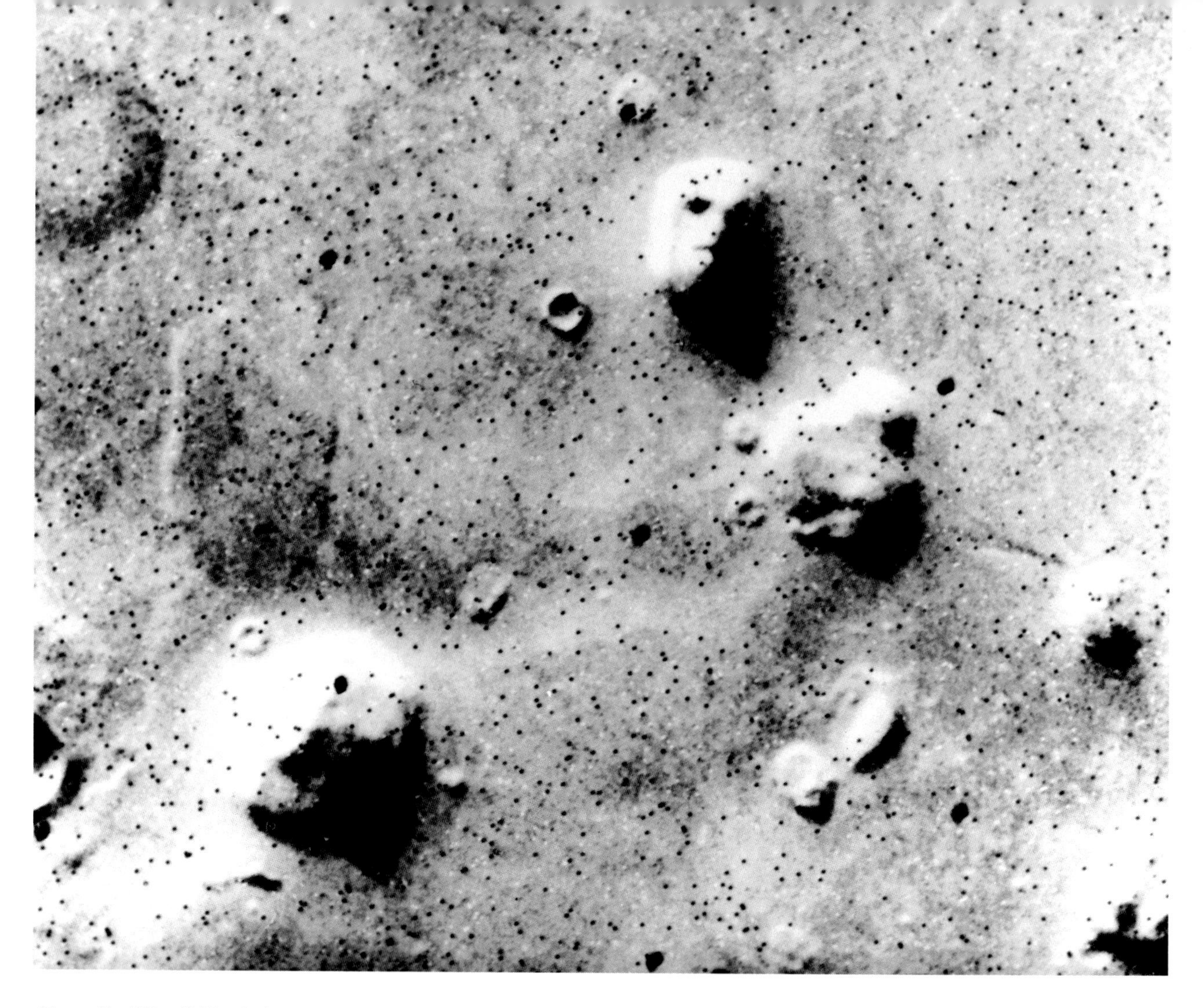

Above: The Viking Orbiter 1 planetary probe was on its 35th orbit around the red planet when its A camera shot frame number 72, in which a monumental humanoid "face" appears. *(NASA)*

some of it flat, some with rock formations and hills rising above a frozen mesa. Cydonia lies 41 degrees north of the Martian equator. Viking dutifully photographed the area and then transmitted its pictures back to Earth.

An imaging team member, Toby Owen, stared in disbelief at Photo Frame 35A72 (taken on the 35th orbit, with the "A" spacecraft, 72nd frame). What he saw didn't seem possible. A helmeted, humanoid face, mouth open, teeth showing, a mile long from helmet top to the bottom of the chin, looked back at him. The "face" loomed a thousand feet above ground level.

Owen turned to another team member. "Oh, my God, look at this!" he exclaimed.

The great debate—acrimonious and angry—was on. To many people who studied the photograph, the image enhanced by computers, some ancient race, either from Mars (unlikely) or from a planet orbiting a distant star (even more unlikely), had created this monolith specifically to be seen from space. A sighting of the "Face of Cydonia," as it came to be called, would be proof positive

Mars had felt the tread of intelligent life-forms and presented this huge structure as an invitation to "come on down."

Most astronomers, scientists, and especially skilled photo interpreters dismissed the "face" as a trick of lighting and shadow. The photo had been taken at 6:00 P.M. Martian time; conditions were perfect for stark relief and shadows. The detractors pointed to the pictures of the moon that "convinced" astronomers that lunar mountains were as jagged and violent in structure as any such features on Earth. That interpretation proved wholly wrong; the mountains of the moon are softly rounded and curved from billions of years of intense solar radiation and extremes of heat and cold that ground them down to nubs.

The argument continues unabated. Natural formation or the signature of a long-vanished intelligent race?

As of the end of 1994, 33 major missions have flown to Mars from the United States and Russia. A flotilla of robot spacecraft from both countries, carrying instruments from a dozen more nations, will set sail during the next few years for landings on the Martian surface.

What they find could be *very* interesting.

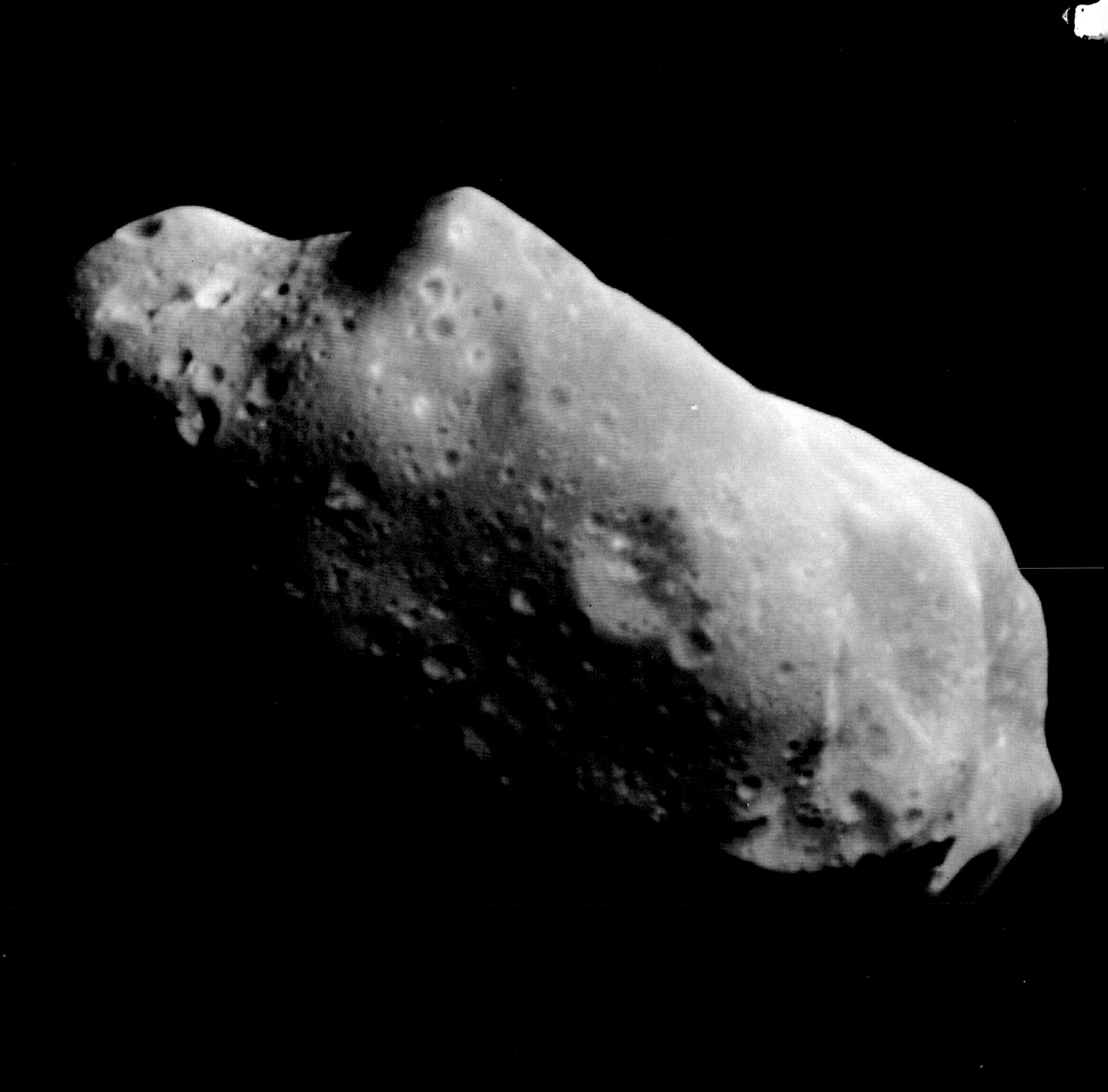

7

THE ROCKY ROAD TO JUPITER

It's the small things in our solar system that hurt the most. Not the planets that orbit the sun, or the 57 known moons that orbit the planets, or even several hundred very small moons that circle large planetoids and asteroids.

What hurts are billions of rock, metal, and frozen ice chunks from the colossal planetary smashups. Almost all this debris swings around the sun like a great necklace of dirty, irregularly shaped pearls known as the Asteroid Belt.

This is the "forbidden zone" of flight outward from Mars to Jupiter. It's an obstacle course 150 million miles deep. That depth and the tumbling, colliding debris by itself aren't the hazard to spacecraft groping through the belt. The thickness from top to bottom is the great trap. A ship flies away from our world along the plane of Earth's orbit. That plane—the ecliptic—may be imagined as a huge sheet of glass for the entire solar system, with almost everything in our system moving along but not above or below that extremely thin passage. Climbing over the Asteroid Belt, or squeezing beneath the grinding obstacles, means climbing or diving at least 25 million miles up or down, and that consumes a staggering quantity of fuel. The only energy-efficient way to traverse the belt is to plunge through its center. That gamble paid off with space probes that sailed through the Asteroid Belt, surviving strikes by micrometeorites and being "sanded" by dust clouds.

But the problem is that not all space debris stays in the Asteroid Belt. Thousands of pieces of the solar system's trash swing in orbits that at times take dead aim at where we live. When the big ones hit, they can pulverize everything where they strike and wreak destruction in every direction for thousands of miles.

For example, the large asteroid that pounded an enormous crater in what is now the Yucatán Peninsula about 65 million years ago wiped out two-thirds of all life on Earth. Other asteroids zing distressingly close to our planet. Another blast like the Yucatán cataclysm means a premature end to civilization.

The really *big* asteroids, like Ceres, with a diameter of 429 miles, are actually small moons without a host planet. The good news is their orbits have been charted, and they are considered "safe" for not aiming at Earth.

But the "wolves of the solar system," as astronomers call certain renegade asteroids, are another question. They wander between the planets, often swinging close to the sun and varying their orbits from year to year. They pass near Earth with tremendous speed and mass; within four million miles of our planet, for example, is considered cutting it close.

Previous page: The large asteroid Ida, like most of its companions in the Asteroid Belt, remains safely distant from Earth. This color picture of the huge, battered chunk of mountain that is more than 35 miles long and 20 miles thick was made by the Galileo spacecraft on August 28, 1993.

To the right of the asteroid one can see that Ida is accompanied by Dactyl, a small, one-mile-wide moon. *(NASA)*

Below: This montage shows asteroid Gaspra, 12 miles long, compared with Deimos (lower left, less than 8 miles across) and Phobos (lower right, with a diameter of 13 miles), the moons of Mars.

The Martian moons Phobos and Deimos were photographed by the Viking spacecraft in 1977. This image of asteroid Gaspra was taken on October 29, 1991, by the Galileo planetary ship. *(NASA)*

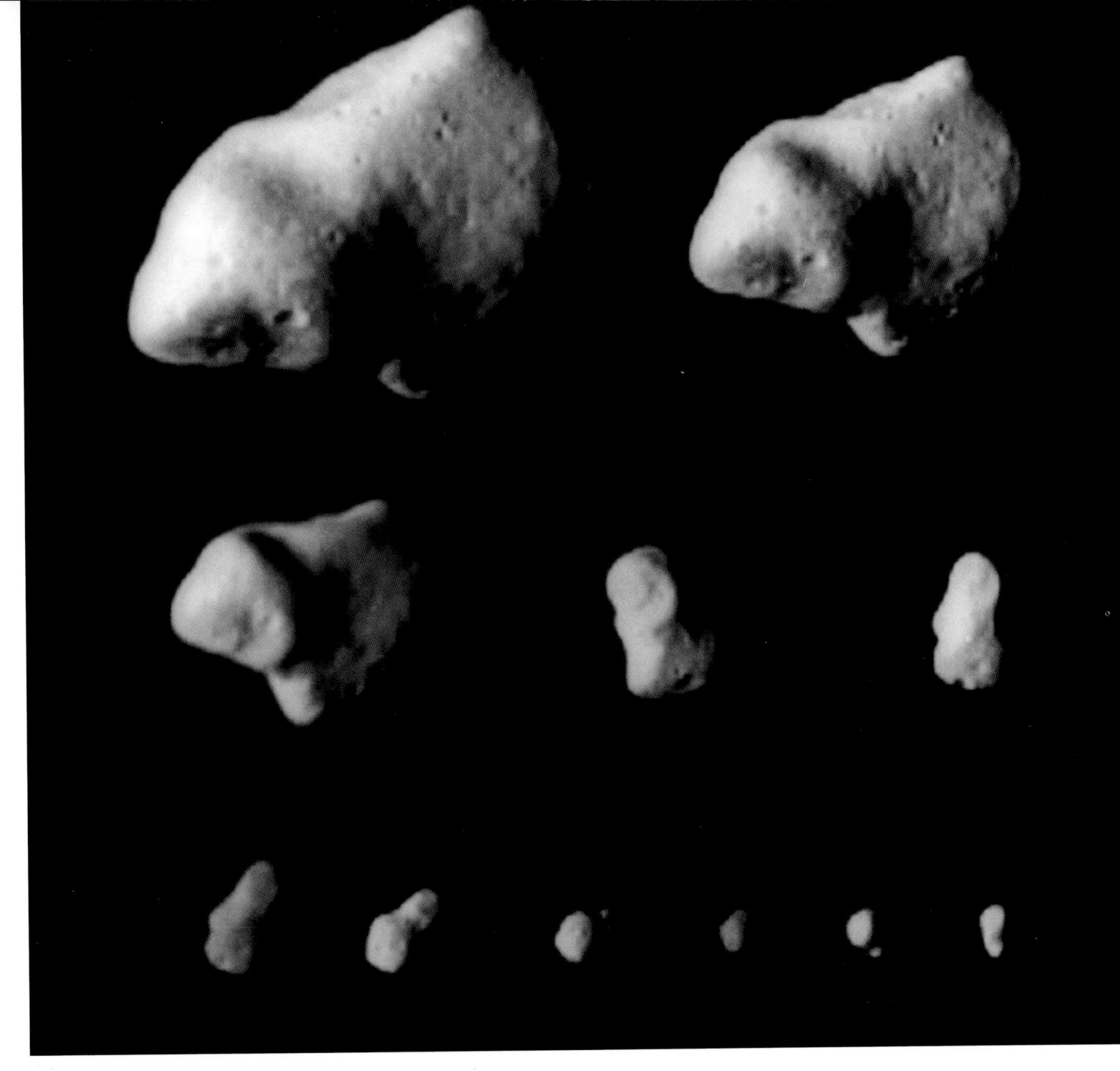

Above: This montage of eleven images of the asteroid Gaspra was taken by the Galileo planetary ship as it flew by the asteroid on October 29, 1991, on its way to Jupiter.

The eleven images show Gaspra growing progressively larger in the lens of Galileo's solid-state imaging camera—beginning with the smallest picture in the lower right and ending with the largest view in the upper left—as the spacecraft approached the asteroid. *(NASA)*

More than 100,000 renegade asteroids prowl the gravitational pathways of the solar system. Unless we make a systematic search for these "celestial wolves," we'll never know when any one or several of the killer asteroids will collide with Earth.

Some near-misses are *scary*. On October 30, 1937, Asteroid Hermes came within 485,000 miles of our planet. If the three-billion-ton Hermes visits again, it could travel a path that would bring it closer than our own moon. Three billion tons slamming into Earth at 70,000 to 100,000 miles per hour is what astronomers call "Lights out."

Until recently, our best instruments perceived asteroids as hazy blobs of light. Even the makeup of comets, masses of ice and rock with sun-lashed pressure blowing dust and gas into huge

Right: The Hubble Space Telescope photographed the comet Shoemaker-Levy and its 710,000-mile-long train as it hurtled at 134,000 miles per hour toward the swirling gaseous ocean of Jupiter.

The dark spot on the disk of Jupiter is the shadow of the planet's inner moon, Io. This volcanic moon appears as an orange and yellow disk to the right of the shadow. *(Space Telescope Science Institute and NASA)*

Below: The comet Shoemaker-Levy had passed near Jupiter on a previous orbit, and in this NASA artist view, one can see how Jupiter's gravity tore the comet into several pieces. *(NASA)*

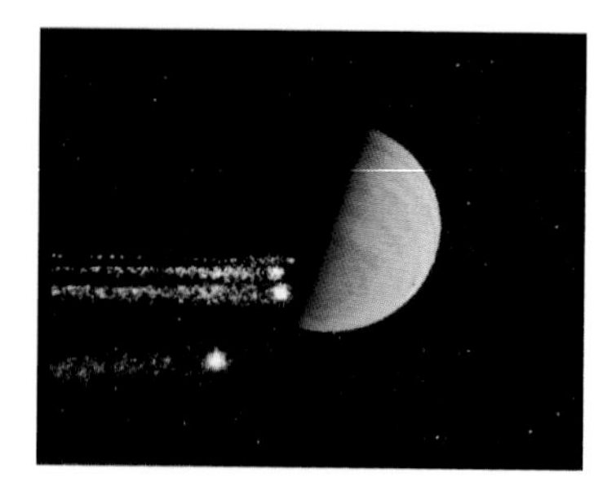

plumes as they neared our star, was more guessed-at than fact-based. Then all this changed. Earth-launched planetary probes dove—literally—into comets, and old concepts of comets as "dirty snowballs" changed to an image of ice-shrouded cores of massive rocks.

Late in 1991, while on a long journey to Jupiter, the Galileo probe made the first close-up visit to the asteroid Gaspra. Two years later, Galileo encountered Asteroid 243 Ida and shot pictures of the stony mountain from a distance of only 1,500 miles. Astronomers discovered Ida had a small traveling companion, Dactyl, a one-mile-wide moon that orbits its host once every 24 hours at a distance of 60 miles. Fortunately, Ida appears to be a permanent member of the Asteroid Belt. A huge, battered chunk of mountain, Ida is more than 35 miles long, 20 miles thick, and bludgeoned with massive craters.

Ida remains safely distant from Earth, but not so Eros, the first asteroid tracked by telescopes away from the Asteroid Belt. One of 200 known asteroids that could potentially crash into Earth, 20 years ago Eros raced by, only 14 million miles from us. In astronomical terms that's agonizingly close, especially knowing that Eros is 22 miles long—about the length of Manhattan Island. If ever Eros plows into our world, it will impact with explosive force measured in gigaton yield—the equivalent of billions of tons of high explosive that could destroy much of the civilized world.

Below: About 65 million years ago, a giant asteroid like the 35-mile-long stony mountain Ida, seen here in this view from the Galileo spacecraft, crashed in the area now known as the Yucatán Peninsula. The collision destroyed most surface life on Earth, including the dinosaurs. *(NASA)*

How big is *big* with asteroids? About 65 millions years ago the rocky nucleus of a comet slammed into Earth at an estimated 70,000 miles per hour. The explosion exceeded 100 million million tons of nitroglycerine. The shock wave tore away from the incandescent fireball at 20,000 miles per hour. The atmosphere itself burned in a vast sheet of flame and incinerated almost everything within its reach.

The blast blew a huge section of atmosphere into space. Moments later, the return pressure wave of the shock front blew back at half the speed at which it left the explosion. Air rushed in to fill the great vacuum existing in the atmosphere, creating a second monstrous turmoil racing around the globe.

Aerial shock waves turned everything into flame for thousands of miles. The planet's crust cracked with huge chasms opening wide for hundreds of feet. Volcanic eruptions followed; seething magma burst upward to flow in a death wave across

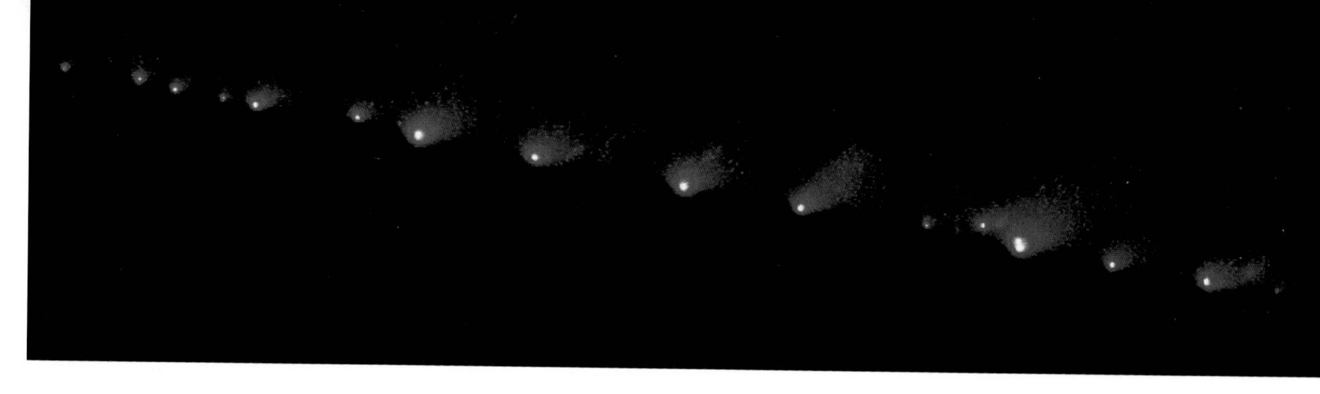

Above: The Hubble Space Telescope took this picture of the comet Shoemaker-Levy's train of 21 icy fragments that stretched across space three times the distance between Earth and the moon. *(Space Telescope Science Institute and NASA)*

much of the world. No one knows how great was the tidal wave; the best estimates say the ocean was hurled apart, boiling and steaming, with billions of tons crashing down on land. Before the fires succumbed for lack of fuel, the atmosphere itself filled with lethal nitric acid.

The asteroid cracked its death whip in the area now known as the Yucatán Peninsula. This ended the reign of the great dinosaurs. Most surface life on Earth whimpered in death throes. Life survived only in the deep oceans far from the blast or in mountain redoubts on the other side of the world.

Eros' "skip" trajectory, as well as even closer near-misses by other massive asteroids, has kicked off a program to track asteroids' orbits and learn how we might deflect one if ever it should be aimed directly at our world.

On March 23, 1989, another asteroid rushed within 700,000 miles of Earth. Much smaller than most asteroids, the 1989 visitor, at a half-mile in diameter, was missed by all Earth telescopes until it was spotted *departing* Earth's atmosphere.

Had the meeting of Earth and asteroid taken place, a small mountain would have smashed into our planet with enough force to kill hundreds of millions of people and destroy cities and countryside over many thousands of square miles.

Jack Hills of the Los Alamos National Laboratory sounds the warning that sooner or later we're going to be hit. For the moment, imagine only a tiny asteroid, 600 feet in diameter, smashing into the Atlantic Ocean far from land. It would generate 700-foot high tidal waves that would race to both the American and European continents. Much of world civilization would be pounded into wreckage in what astronomers judge would be the greatest catastrophe in all of human history.

Comet Shoemaker-Levy was the star cosmic attraction of 1994 when it hurtled at 134,000 miles per hour into the swirling gaseous ocean of Jupiter.

The Mount Palomar Observatory first sighted Shoemaker-Levy on March 24, 1993. Astronomers noted that what had once been a solid core with outlying ice and dust had passed near Jupiter on a previous orbit. The Jovian giant gravity tore the comet into a long strand of individual rock particles like glowing diamonds. The comet was drawn into intercepting Jupiter—its final destruction assured by powerful gravity drawing the comet ever closer. On July 21, 1994, scientists observed for the first time the impact of one orbiting body into another. Shoemaker-Levy slammed like rapid-fire cannon shots into the gaseous mantle of Jupiter. The cannonade proved an unprecedented fireworks display. Before breaking up into the string of blazing diamonds, the original comet was six miles in diameter and weighed *500 billion* tons.

Fireballs erupted from Jupiter as the huge rocks ignited almost instantly from friction with the upper clouds of ammonia ice. For six days the bombardment continued. At one point a mushrooming ball of superheated gas roared upward from Jupiter, 50 times brighter than the entire planet, blinding telescopes on Earth. By the time the last pieces of comet had plunged into Jupiter, the friction explosions released explosive energy equal to six million million tons of TNT. Black smudges,

Below: Comet Shoemaker-Levy made a close approach to Jupiter in the summer of 1992 and was broken into multiple pieces by the force of gravity from the giant planet, as seen in the image on the left made on March 30, 1993, with the Spacewatch Camera of the University of Arizona.

Center image: On July 1, 1993, when the comet's "String of Pearls" was 100,000 miles out, the Hubble Space Telescope shot this view of Shoemaker-Levy heading for Jupiter.

The image on the right, taken by Hubble when the comet was 40,000 miles from Jupiter, shows the brightest nucleus of the "String of Pearls." *(University of Arizona and NASA)*

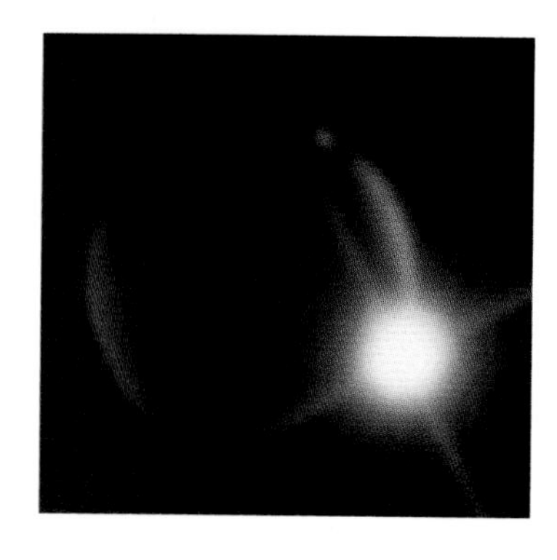

Above: Comet Shoemaker-Levy bombarded Jupiter for six days. At one point, as seen in this photograph made 12 minutes after impact of Fragment G by Peter McGregor ANU 2.3m telescope at Siding Spring, an explosion 50 times brighter than the entire planet blinded telescopes on Earth. *(Peter McGregor ANU 2.3m telescope at Siding Spring and NASA)*

Below: The eight black smudges on the Jovian surface, seen here in a photo taken by Hubble, were testimony to the violence caused by comet Shoemaker-Levy. *(Space Telescope Science Institute and NASA)*

each larger than our planet, on Jupiter's surface were testimony to the violence now past.

No one questioned that had this comet targeted Earth, virtually all life would have perished.

Nearly every observatory in the world followed that death ride to Jupiter. Six spacecraft trained their instruments on the dazzling celestial fireworks. The Hubble Space Telescope team took advantage of the rare opportunity to take more than 400 photographs of the blasts.

The bruising of the Jovian atmosphere sounded a clarion call for tracking every possible interloper that might one day strike this planet. At a NASA symposium in Washington, Professor Carl Sagan estimated there was once chance in a thousand that in the next century a major comet or asteroid would blast Earth. Sagan emphasized these were not very good odds. He told the symposium attendees and the press that "you would not go on a commercial airliner if the chance of it falling were one in one thousand."

Political action for identifying space debris orbits came immediately—including maximum use of Hubble's superior viewing powers to begin the earth-protection effort. Jupiter was still sporting massive bruises when NASA formed the Near-Earth Object Search Committee, with the goal of detecting potential killer

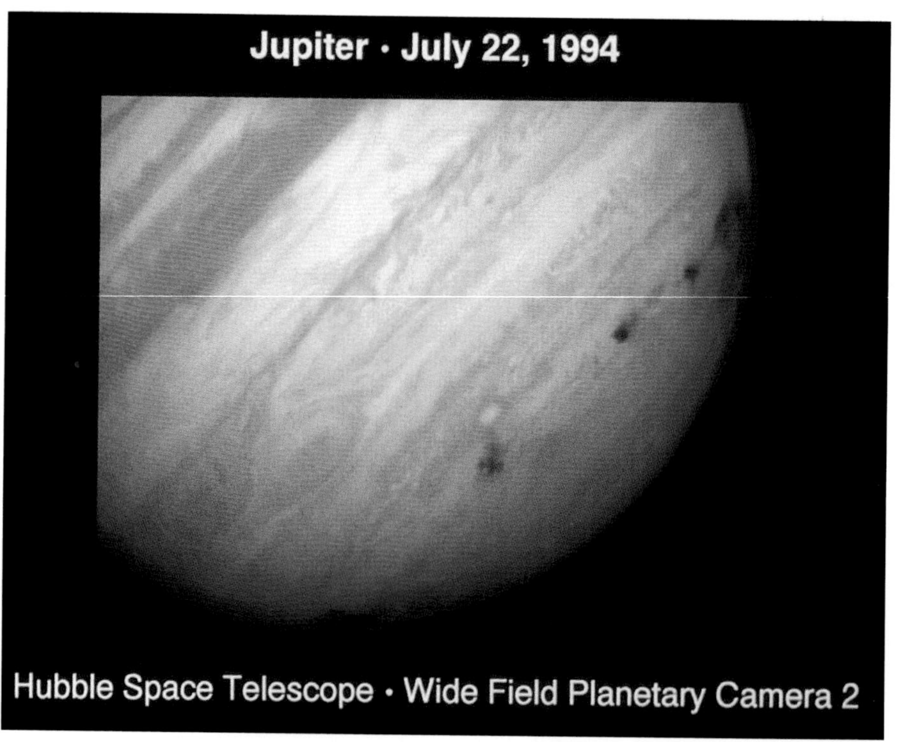

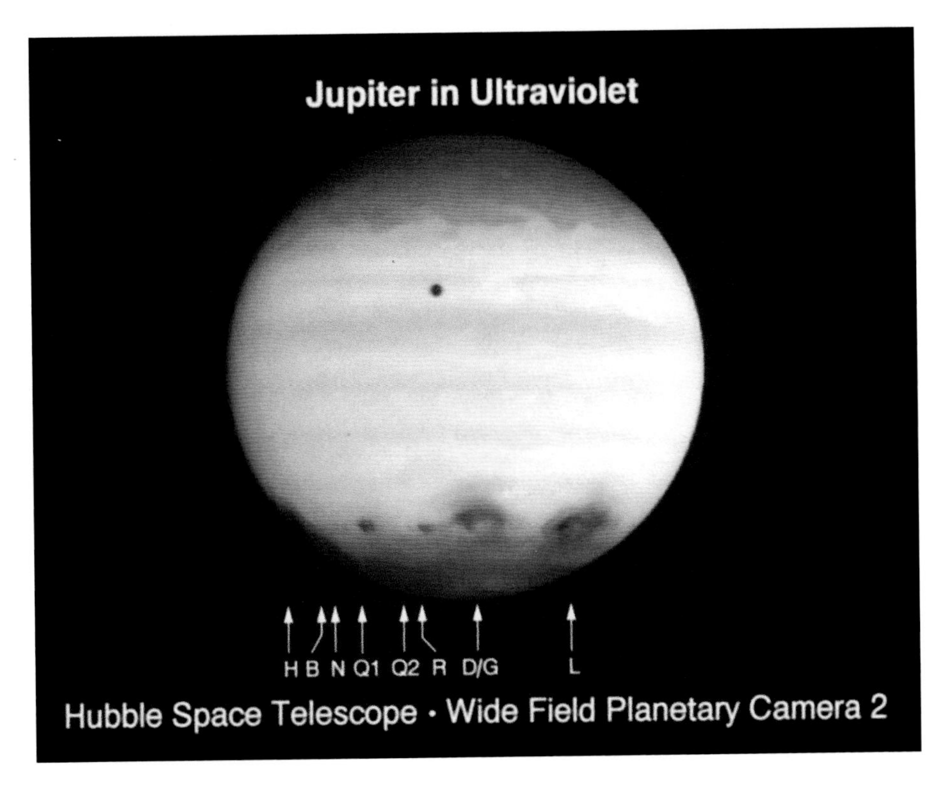

Left: The bruising impacts on the Jovian atmosphere are cataloged here in this ultraviolet picture. The comet Shoemaker-Levy's collision with Jupiter led NASA to form the Near-Earth Object Search Committee to detect the approach of killer asteroids and comets. *(Space Telescope Science Institute and NASA)*

asteroids and comets, and devising a means of either exploding them or deflecting their approach to Earth with thermonuclear weapons.

Astronomer Eugene Shoemaker and the NASA group warned that there are at least 15 comets known to swing within Earth's orbit that could wreak global devastation if they struck this planet.

Leading astronomers warned Congress that each day an asteroid the size of a house passes between the Earth and the moon; each month, one the size of a football field passes between them.

Hubble, demanded national leaders, must join the long-term effort to track the "*wolves of the solar system.*" NASA scientist Donald Yeomans stressed that every step possible should be made to prepare for an oncoming asteroid strike. "When I started studying the motions of comets and asteroids in the early 1970s," he stated, "we only knew about two dozen near-Earth asteroids. . . . Now we're up to two hundred fifty, and we're still looking *for another twenty-two hundred.*"

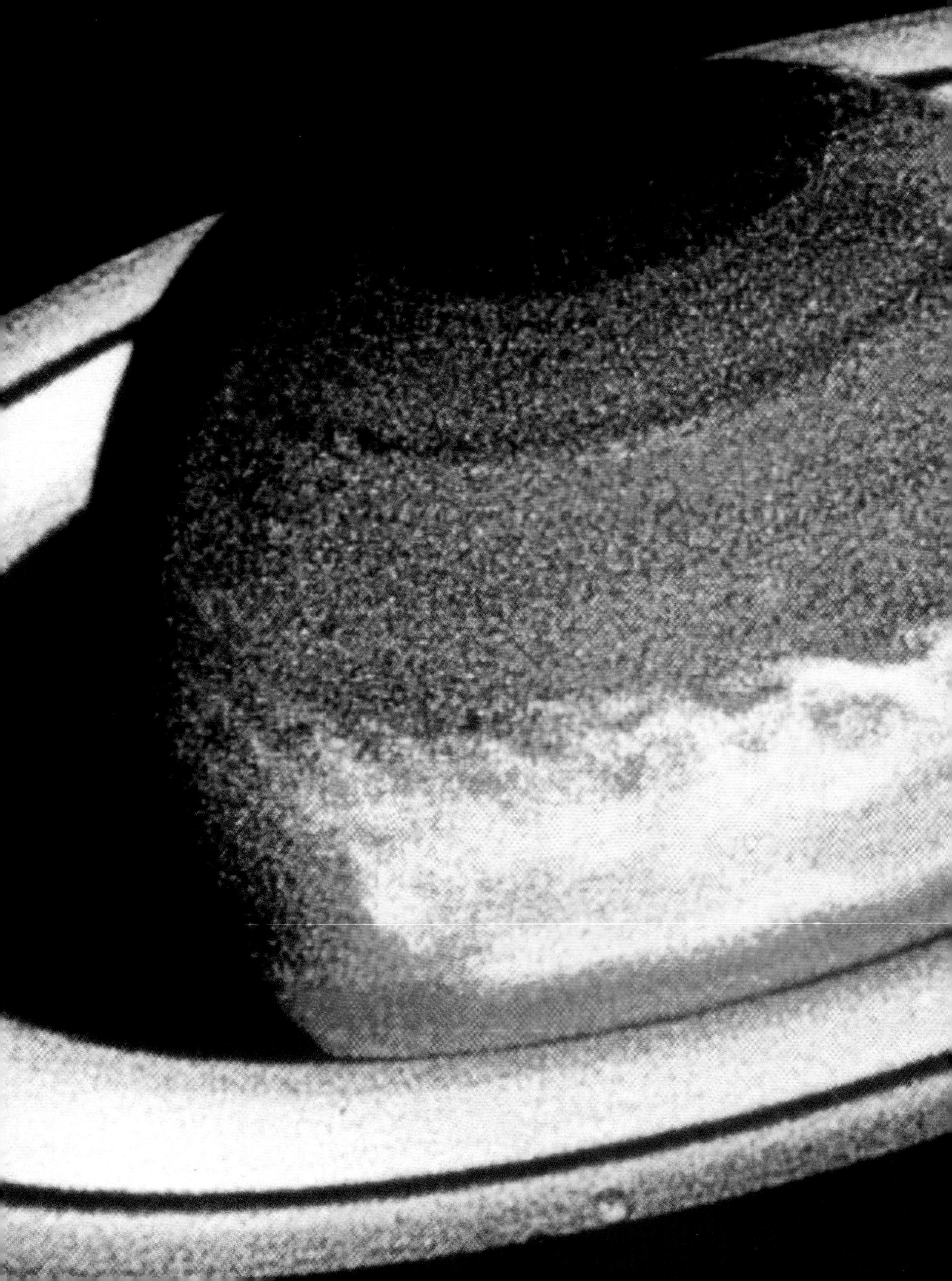

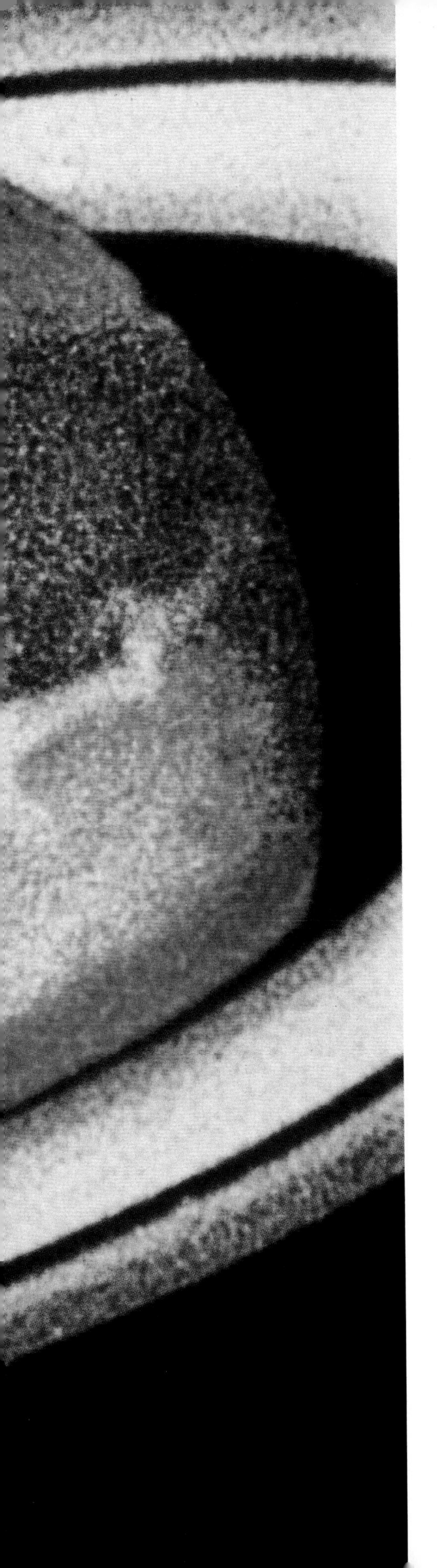

8 THE GAS GIANTS

Nearly 400 years ago Galileo Galilei assembled a small telescope with crude lenses. With his invention, he banished forever a past of ignorance and opened the real heavens to humankind.

That first night when Galileo turned his telescope to the light-spattered sky, he began the long road that would lead ultimately to the Hubble Space Telescope. Time, size, power, and technology are the gulf between Galileo and Hubble, but the purpose and wonder, the leap across unimaginable voids, are shared equally.

Overjoyed and humbled, Galileo stripped away the ignorance shrouding other planets. He quickly learned the different phases of gleaming Venus. That this shining light in the night sky was a world rolling obediently about the sun was proof positive of worlds other than Earth, proof that the sun, with some invisible force, had gathered a family of globes.

Other worlds beckoned; Galileo sought the moon. He saw through chromatic discoloration of light caused by his improperly ground lenses. No matter! Again came awestruck revelation. Until this very moment the moon had been a featureless, bland globe.

Previous page: As seen in this photo taken by the Hubble Space Telescope, Saturn is streaked by violent weather and high winds that reach speeds of more than 1,100 miles per hour. *(NASA)*

Right: Galileo's first deep look into the night must have been akin to this magnificent view of Jupiter rising above the airglow over Earth's horizon, as seen from the space shuttle Endeavour. The crescent moon is at top with the giant planet shining brilliantly below. *(NASA)*

Now Galileo stared in wonder at what appeared to be great mountain ranges, sawtooth in profile against blackness. Huge craters, enormous geographical rays, tumbled terrain leaped into his view.

Astonished at a world that changed completely with a single sweeping glance, he turned his telescope on another distant globule of light. Jupiter swam into view and Galileo knew he was now opening magnificent vistas for man. Through his lenses Jupiter changed from that dewdrop of light to a great swollen world. But whatever the colossus offered, it paled before the stunning glimpse of four tiny shining motes.

Those sparks etched against blackness . . . moons of Jupiter! That ended the past and began the future. If smaller worlds revolved about larger host planets, then the cosmology to which church and science adhered was disastrously *wrong*. Until this moment, Earth was believed to be the center of the universe with everything else revolving about it.

For another century, the rulers of nations and men railed against Galileo and his adherents. One attack upon the astronomer spoke for them all—a condemnation from the Aristotelian professors who ranked as the contemporaries of Galileo: Jupiter's moons are invisible to the naked eye, and therefore can have no influence on Earth, and therefore would be useless, and therefore do not exist.

The early telescopes didn't help much to properly define the next gas giant beyond Jupiter. Galileo judged the rings of Saturn as anything *but* rings. The limited focus of his telescope first led him to believe the rings were actually two moons. Continued observa-

Below: Early astronomers were fascinated with Jupiter, but whatever the Jovian colossus offered, they were stunned by the glimpse of the 4 largest of Jupiter's 16 moons, seen in this NASA picture showing the giant planet with its moons Io, Europa, Ganymede, and Callisto. *(NASA)*

Above: Jupiter is 11.2 times the diameter of Earth, at 88,900 miles, and has 16 known satellites. Two Jovian moons, Ganymede and Europa, can be seen with Jupiter in this photograph taken by Voyager. *(NASA)*

tion brought him to dismiss Saturnian moons and conclude that what he saw were *handles* on each side of the planet, as if it were in the shape of an enormous rounded soup mug.

As telescopes increased in power and became able to focus on details once too small even to be detected, Jupiter became a popular object of study. The largest planet of the solar system was banded by strange colors and streaks that whirled constantly. A huge anticyclone, the Great Red Spot, baffled observers until they were able to define the oval area as a storm that had raged for hundreds, perhaps thousands of years, and was more mystery than explainable. More powerful telescopes confirmed the giant planet was, at 88,900 miles, 11.2 times the diameter of Earth, and was also a miniature solar system unto itself, with 16 known satellites.

Modern astronomy scanned Jupiter with instruments that discern more than the visible light collected by optical telescopes. Today the term *telescope*, especially with the powerful Hubble, refers to an instrument that is able to "see"—sense—across a wide band of the electromagnetic spectrum. The modern array of scientific instruments study distant planets, moons, stars, gas clouds—anything visible *and invisible*—by spectroscope, radiation detectors, mass detectors, and other equipment that seem almost magical as they garner every detail possible. What these instruments present is much like the information bands that lasers scan in grocery and department stores. On a wide band, visible light is but a single dark line, the width of a human hair in an inch-wide band. Other data are gathered in ultraviolet and infrared light, measurements of radiation, gravitational effects—the widest possible signature of the cosmic target. The secret to this remarkable science is that spectroscopic analysis proved that *every* chemical element can be identified by its spectral identity code. In its chemical makeup, hydrogen is the same on Earth, in the sun, in the at-

Above: The giant of all planets, Jupiter, seen in a photograph taken by the Hubble Space Telescope on May 28, 1991, before its focus problems were corrected by space-walking astronauts. *(NASA)*

Below: The Great Red Spot, a huge anticyclone on Jupiter's surface, has raged for hundreds, perhaps thousands, of years. *(NASA)*

mospheres of other planets, and as far back in time and distance as even Hubble can gather light. Whether distant galaxies, tenuous multicolored gas clouds wisping through space, or the swirling clouds of Jupiter, each can be measured as to its chemistry, motions, and the pull of gravity on its surface.

What is invisible becomes kaleidoscopic variety in a wide band of energy other than visible light.

Below: The powerful Hubble Space Telescope scans Jupiter with more than visible light gathered by its mirrors. Its array of instruments and other equipment is able to study cloud movement on Jupiter close up. *(NASA)*

Jupiter breaks the rules; it doesn't behave like any other planet. Its mass is 318 times that of Earth, yet it is only one-fourth as dense. Earth completes a rotation once every 24 hours; Jupiter, with a diameter 11.2 times greater, spins about completely every 9 hours 50 minutes and 30 seconds. Along the equator, Earth spins at

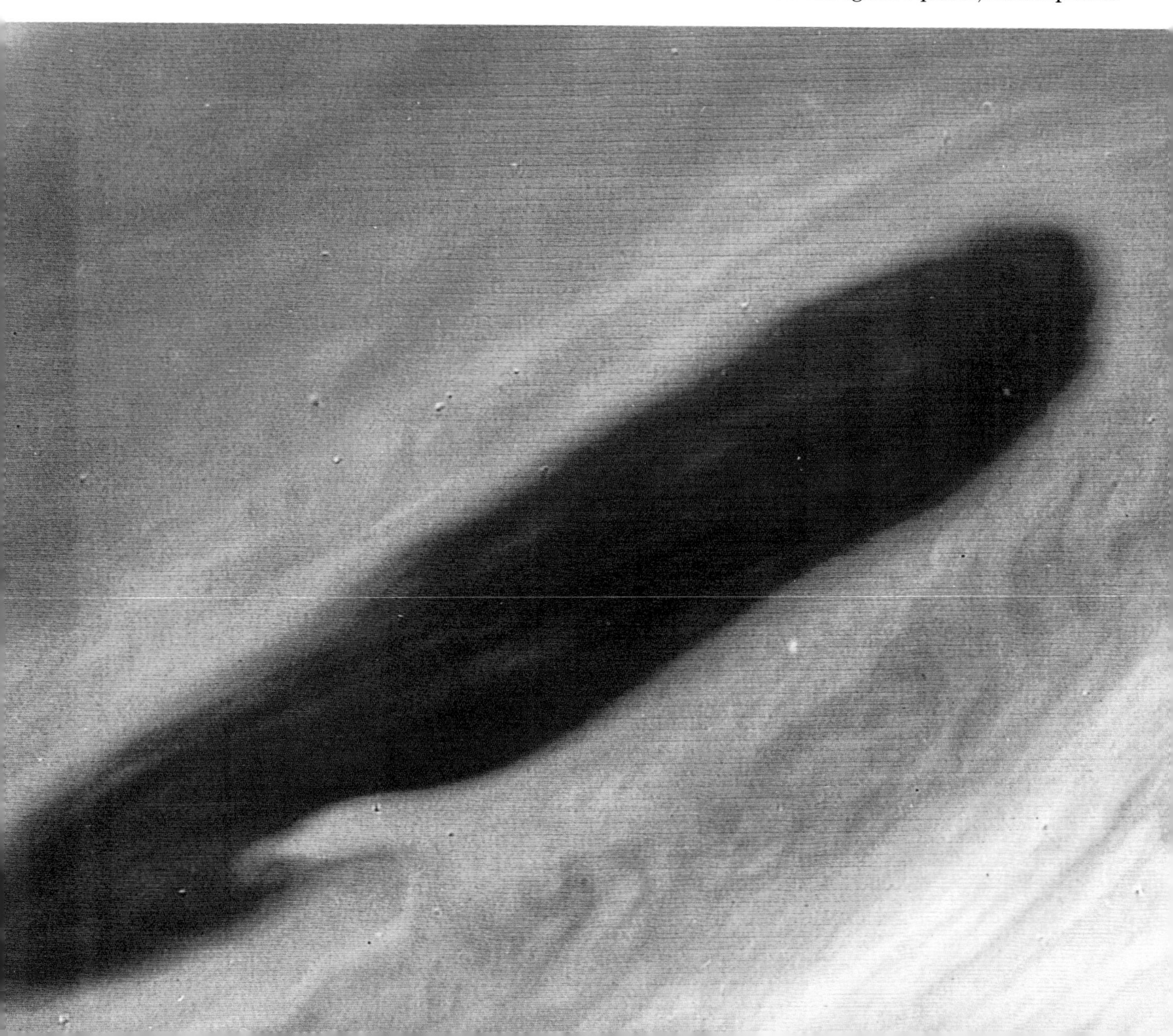

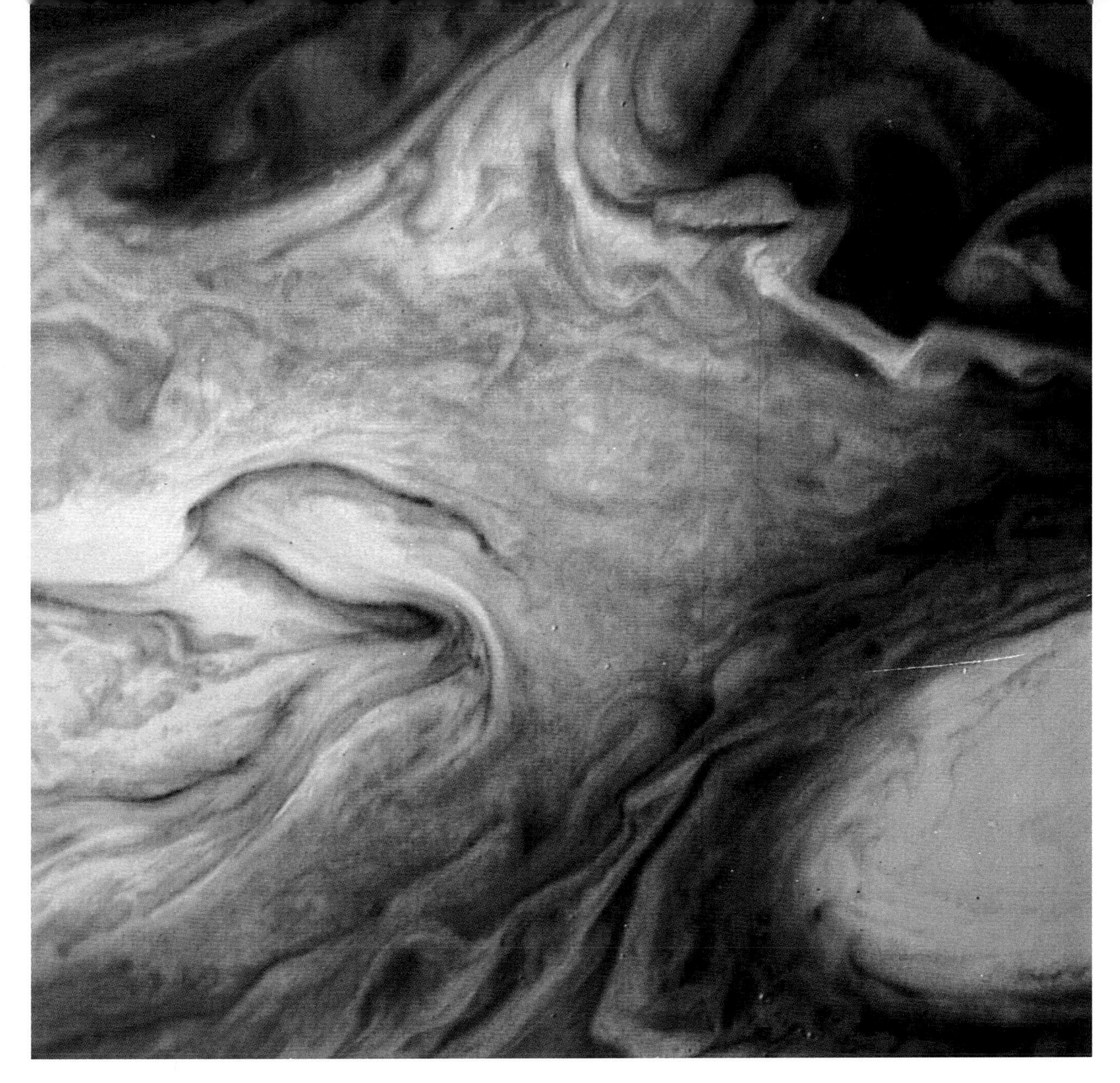

Above: The perilous surface of Jupiter, as photographed by Voyager. Astronauts in spacesuits (or even their spaceship) entering Jupiter's atmosphere would be on a one-way road to disaster. *(NASA)*

just under a 1,000 miles per hour. Jupiter rotates at 22,000 miles per hour, a whirling motion so powerful the planet is visibly flattened at its poles and bulges along the equator.

Jupiter's orbit about the sun averages 480 million miles from our star, far enough to reduce the radiation energy falling on Jupiter to less than four percent of the warmth we receive. Jupiter falls clearly within the range of a "cold planet," yet the planet radiates *away* twice as much heat as it gets from the sun.

Astronomers explain this heat radiation as the result of the enormous world compressing against its small, rocky core. Deep into the below-surface structure of the planet, hydrogen and helium are compressed to liquid heavier than metal.

This compression leads scientists to conclude that Jupiter is a "failed star," a world amassing material to just below that point

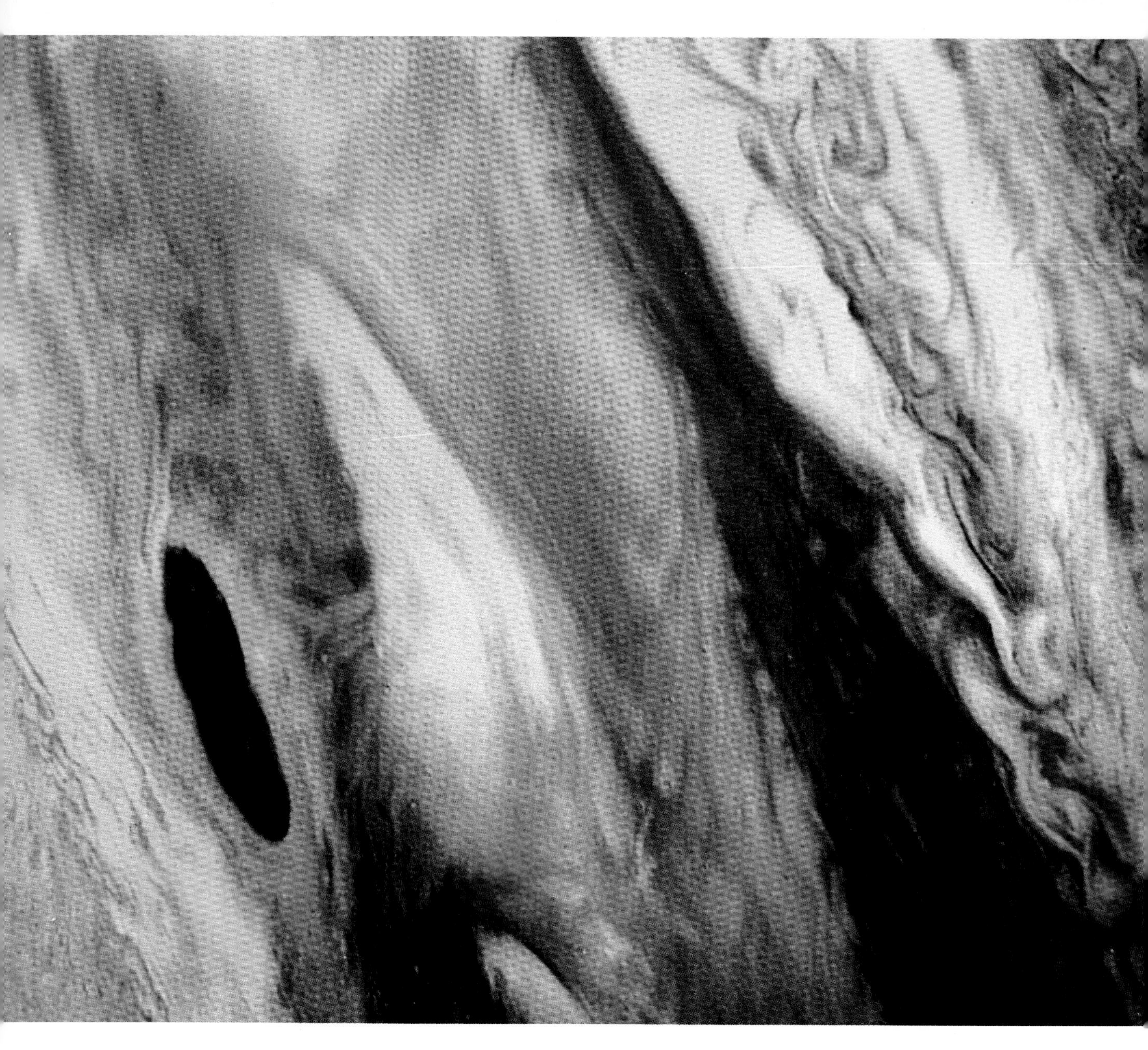

Above: The Voyager probes took this and thousands of other pictures, and recorded vast quantities of data from measuring instruments. They found the Jovian winds were blowing at over 300 miles per hour. *(NASA)*

where the core might ignite spontaneously. "If [Jupiter] were only a little more massive," explains Dr. Tobias Owen of the Illinois Institute of Technology Research, "gravitational contraction would release so much energy that it would turn into a nuclear furnace, like the sun or any other star, and become incandescent."

Comparing Jupiter's surface conditions to those of Earth provides a critical yardstick for Hubble and other deep-space searchers. One major measuring rod with which to describe Earth's surface is "mean sea level," an average point from which we measure height above (as in mountains) or depth below (as beneath the oceans or as in some land features—Death Valley and the Dead Sea). We always have that safe, comfortable, available reference point.

Earth has an atmosphere distinct from anything else above ground or ocean level. There is either water or hard surface; everything above that surface is atmosphere. Looking for a surface on Jupiter is a losing and lethal endeavor. The surface of Jupiter is strictly a *theoretical* level. Immediately upon penetrating Jupiter's atmosphere, a visitor would slip into poisonous gases tumbling violently. Visibility would vanish in the turbulent "air" of ammonia ice. Descent would continue into the main atmosphere of hydrogen and helium, splashed with lesser amounts of methane, ammonia, ethane, acetylene, phosphine, germanium tetrahydride, and hints of ammonia and water crystals. At times the temperature is bitterly cold, then it increases swiftly. Two hundred and fifty miles down, the temperature soars to 1,400°F.

But that wouldn't matter to the descending visitor. Gravity presses against the heavy gases, which then become liquid much heavier than metal, able to conduct electrical charges. The visitor would be crushed by pressure and consumed by enormous electrical current.

Scientists knew that winds blow fiercely on Jupiter, suspecting speeds up to 100 miles per hour. Then the Voyager probes swept past the planet, taking 33,000 pictures and amassing immense reams of data. They found Jovian jet streams thundering along at better than 300 miles per hour. How and why is still a mystery to be solved. Said Bradford Smith, astronomer at the University of Arizona and member of the Voyager team, "The existing atmospheric circulation models have all been shot to hell by Voyager."

Below: In this picture of sunrise on Jupiter taken by Voyager, a thin, dusty ring can be seen glowing in a straight line away from the Jovian surface. *(NASA)*

Below: Astronomers found the best way to see Jupiter's ring was with infrared lenses. The Jovian ring is shown clearly here in this infrared picture made by the Keck Observatory in Hawaii. *(Keck Observatory)*

Enter the growing demand for training the Hubble Space Telescope on the giant gas planets, especially Jupiter and Saturn. Space probes like Voyager performed brilliantly to create an in-depth portrait of such worlds. But they're there briefly, and then they're gone. Hubble can scan Jupiter at regular intervals and amass *consistent* patterns and changes of the surface. Astronomer Reta Beebe of New Mexico State University said, "Planetary people will do anything they can for more time on the Hubble. We aren't even shameful about it."

After all, there are 24 billion square miles of planet surface to survey.

As pictures from passing deep-space probes returned to Earth, Jupiter seemed limitless in its surprises. Until brief but spectacularly successful photographic passes, Saturn was known

Above: Jupiter's moon Ganymede is the largest of all planetary moons. Seen here in a Voyager photograph, Ganymede is larger than the planet Mercury. *(NASA)*

to be circled by immense rings—which led astronomers to suspect that other gas giants also would wear a halo of ice, dust, and rock. They were right: Thin, dusty rings also circle Jupiter.

Jovian moons appeared as if conjured by magic. Sixteen have been photographed, some only 25 miles across. But the kingpin of all planetary moons proved to be Ganymede; its diameter of 3,157 miles exceeds even the size of the planet Mercury. Ganymede is slated for intensive study and exploration. It is a startling world with typical fractures, craters, and rough terrain, but it is also half water ice and half heavy rock material. Its fame comes from confirmation of frozen oxygen on the surface, the first such discovery on any dense planet or moon other than our world.

Above: The Jovian moon Ganymede, as seen here in this Voyager photo taken from a distance of 151,800 miles. *(NASA)*

Europa and Callisto provide startling contrasts for the moons of the same planetary system. Europa has a surface of frozen water ice that resembles a melting country lake with fractures zigzagging its surface. Does that water ice offer hope for life-forms? Perhaps, and thus Europa becomes a candidate for intensive future scrutiny. Callisto is a stepchild of the Jovian system, bigger than Earth's moon but with a surface so smashed and cratered it makes Earth's moon a smooth desert by comparison.

And finally there is the netherworld of Io, fiery with constant eruptions. At 2,255 miles in diameter, Io is remarkably similar in size to our moon. All resemblance ends there, for Io is so torn by gravity-driven tidal forces from Jupiter that it is literally a volcanic world. During the Voyager flybys, at least nine powerfully active volcanoes were photographed, another was almost certainly intermittently explosive, and it seemed clear that hundreds of other volcanoes erupt sporadically. Io is a world that every few hundred years seems to turn itself inside out—and in the process this one moon affects the entire Jovian system.

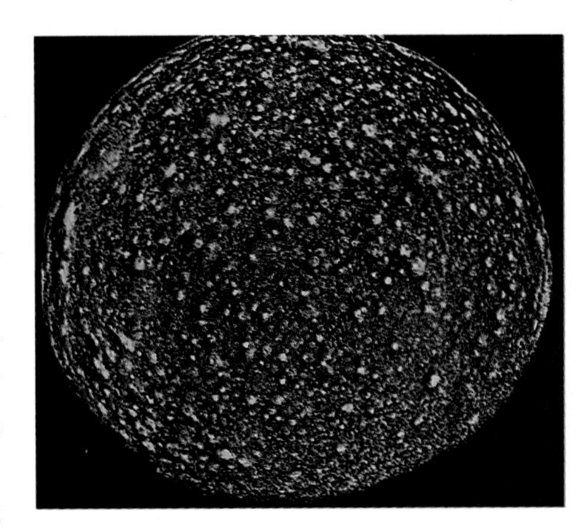

Above: Callisto, as seen in this Voyager photograph taken about the same distance in which we view our moon (245,000 miles). *(NASA)*

Below: The Jovian moon Europa, as seen in this Voyager close-up from a distance of 150,000 miles. *(NASA)*

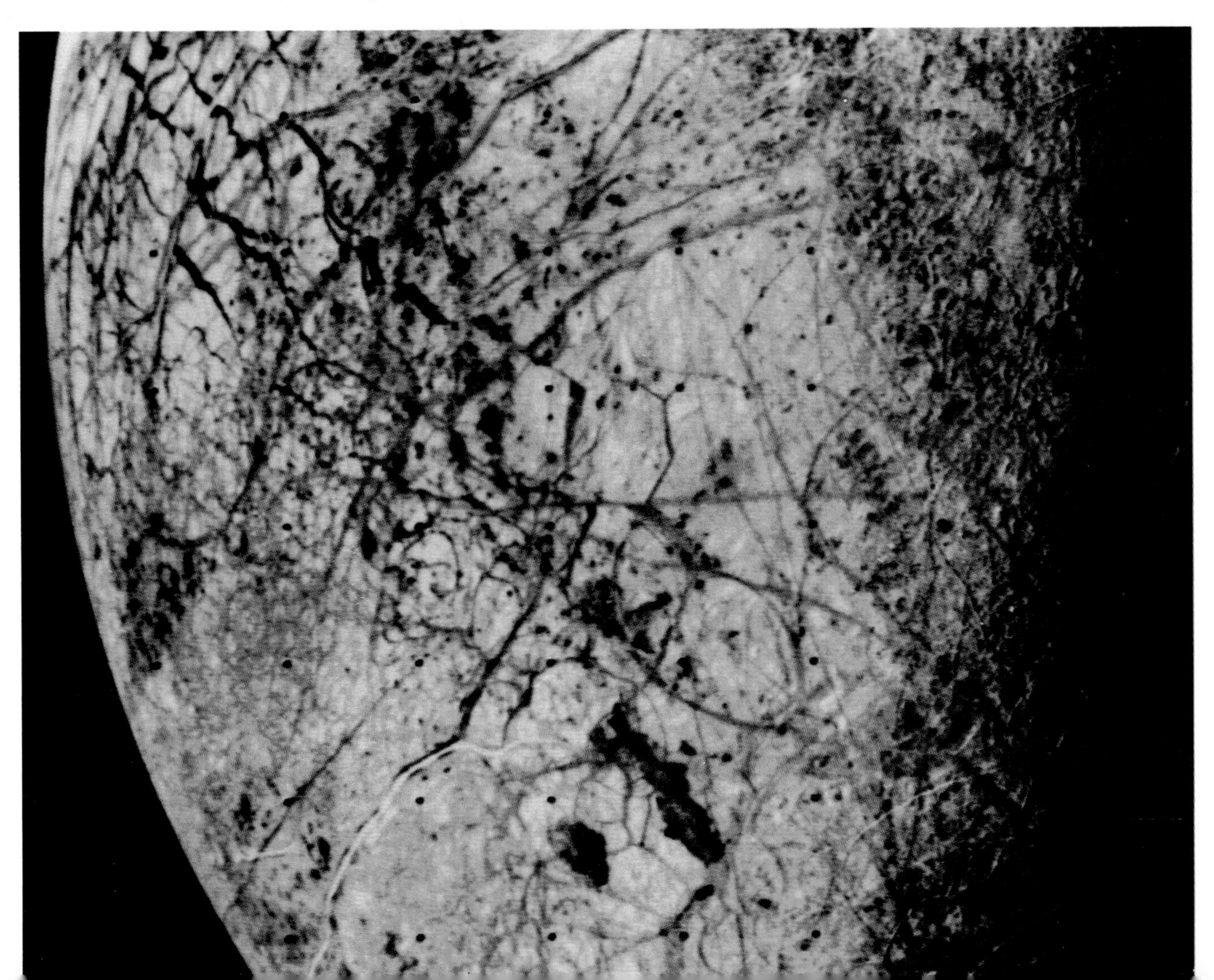

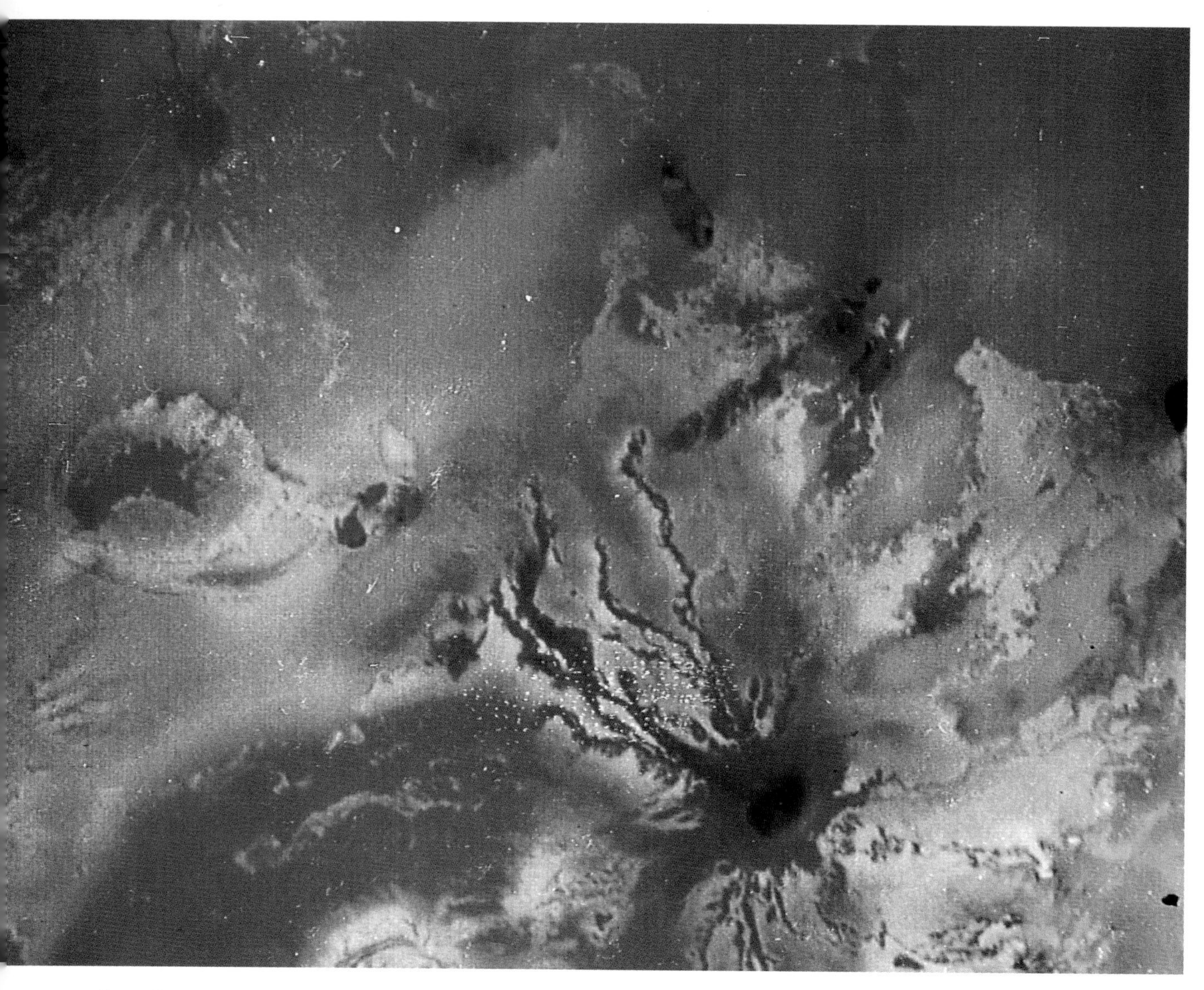

Above: The volcanic surface of Io can be seen in this Voyager photograph. *(NASA)*

Lightning is a constant, booming menace on Jupiter, thunder blasting and echoing for thousands of miles from bolts the size of terrestrial rivers. But there is another force slashing upward from the huge world, an interplay of electrical energies between planet and moon. Io ejects from its volcanoes huge clouds of sulfur, oxygen, and sodium at 20,000 miles per *second*.

The tremendous magnetic field of Jupiter draws in these emissions as downward-spiraling electrical tornadoes. Brilliant auroral patterns flash from Io to Jupiter in spectacular colors and changing intensities, then race around the planet in shimmering waves. This is only the beginning of the electrical charges building between the two globes. When they reach their peak, the electric current boosts to more than five million amperes, a flux tube of explosive energy struggling for release.

It comes finally in a supercharged titanic blast, lightning bolts miles wide, crashing across space until Jupiter and Io are joined in a clash of electrical fire.

Jupiter's solar-system-within-a-solar-system is invaluable to the searches for other solar systems well under way by Hubble and astronomical observatories on Earth. First results of scanning other stars indicate large dark bodies orbiting some stars. Measurement by visible light and across the entire spectrum of energy consistently shows that what have so far been observed are huge planetary forms of mass and size comparable to Jupiter and the other gas giants of our own solar system.

Jupiter has become the yardstick with which Hubble will expand its search.

Below: The massive volcano seen here on Io's horizon is ejecting huge clouds of sulfur, oxygen, and sodium at 20,000 miles per second 100 miles upward. *(NASA)*

Above: Saturn, as photographed by Voyager. *(NASA)*

Below: A Voyager photograph of Saturn's northern hemisphere. *(NASA)*

WORLD OF MAGIC RINGS

The sixth planet from the sun is Saturn, a freak among worlds. It's a massive globe of spectacular color, girdled by thousands of multihued rings, and as planets go, it's so light it might as well be made of cotton candy. In fact, if we could find a lake large enough to hold the 75,000-mile-wide gas giant, it would float like a cork.

But what Saturn lacks in density it makes up for in its gossamer substance as the most beautiful world in all the solar system. It is a globe of subdued pastels; a butterscotch hue dominates. Like Jupiter, Saturn's surface shows bands and jet streams of violent weather storms, turbulent clouds, and winds of more than 1,100 miles per hour.

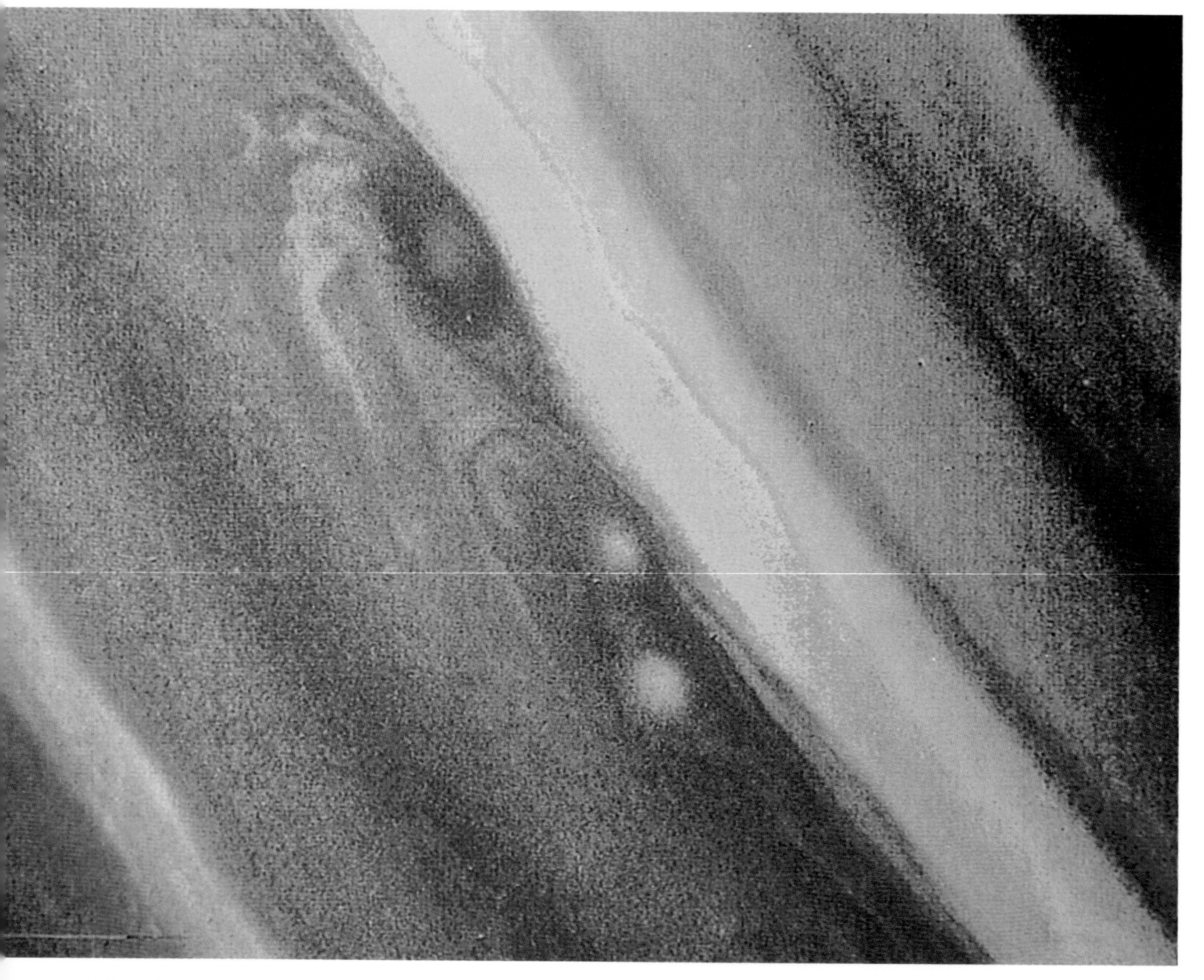

Above: Like Jupiter, Saturn is composed primarily of hydrogen and helium. *(NASA)*

Saturn mimics the composition of its larger, heavier neighbor as well, being primarily of hydrogen and helium. Its great rotational speed of just over 10 hours and 19 minutes, again like Jupiter, flattens the poles, bulges the equator, and creates long-lived oval storms. In many ways, Saturn is just like Jupiter on a slightly less scale, with two exceptions.

At least 20 known moons orbit the huge globe. Water ice forms the major material of Mimas, Enceladus, Tethys, Dione, and Rhea, an astonishing find for moons so far from the sun that they receive but one percent of the radiant energy falling on Earth. Enceladus especially is a jewel of unusually pure water ice; it reflects almost 100 percent of the sunlight striking the globe.

Those moons are interesting, but one moon above all fascinates scientists of all disciplines for what may be hidden beneath its thick atmosphere. At 3,200 miles in diameter, Titan is a thousand miles wider than Earth's moon, but Titan is the *only* moon in all the solar system that has an atmosphere—and may harbor primitive life-forms.

Above: Saturn can be seen in this NASA photograph, with 6 of its 20 known moons. *(NASA)*

This planet-sized moon of glowing reddish haze may be the time machine the Hubble team is looking for. It may be the time capsule to reveal Earth as it was when life-forms first appeared on our world. "For almost two decades," states a NASA report on Titan, "space scientists have searched for clues to the primeval Earth. The chemistry going on in Titan's atmosphere may be similar to that which occurred in the Earth's atmosphere several billion years ago. . . . Atmospheric pressure near Titan's surface is . . . 60 percent greater than Earth's. The atmosphere is mostly nitrogen, also the major constituent of Earth's atmosphere. . . . Titan's methane, through continuing photochemistry [when combined with nitrogen] becomes hydrogen cyanide . . . an especially important molecule, since it is a building block of amino acids."

There's a missing link in confirming the nature of Earth's atmosphere billions of years past. Then, methane, ammonia, water vapor, and hydrogen—certainly not oxygen—covered the planet. Laboratory experiments show that when hit by lightning, this mixture formed new material rich in complex organic compounds, which could have fed primitive organisms crawling upward from the oceans. Hundreds of various forms of bacteria attacked the new food source, known as tholin.

Below: Saturn's moon Enceladus is only 310 miles in diameter. *(NASA)*

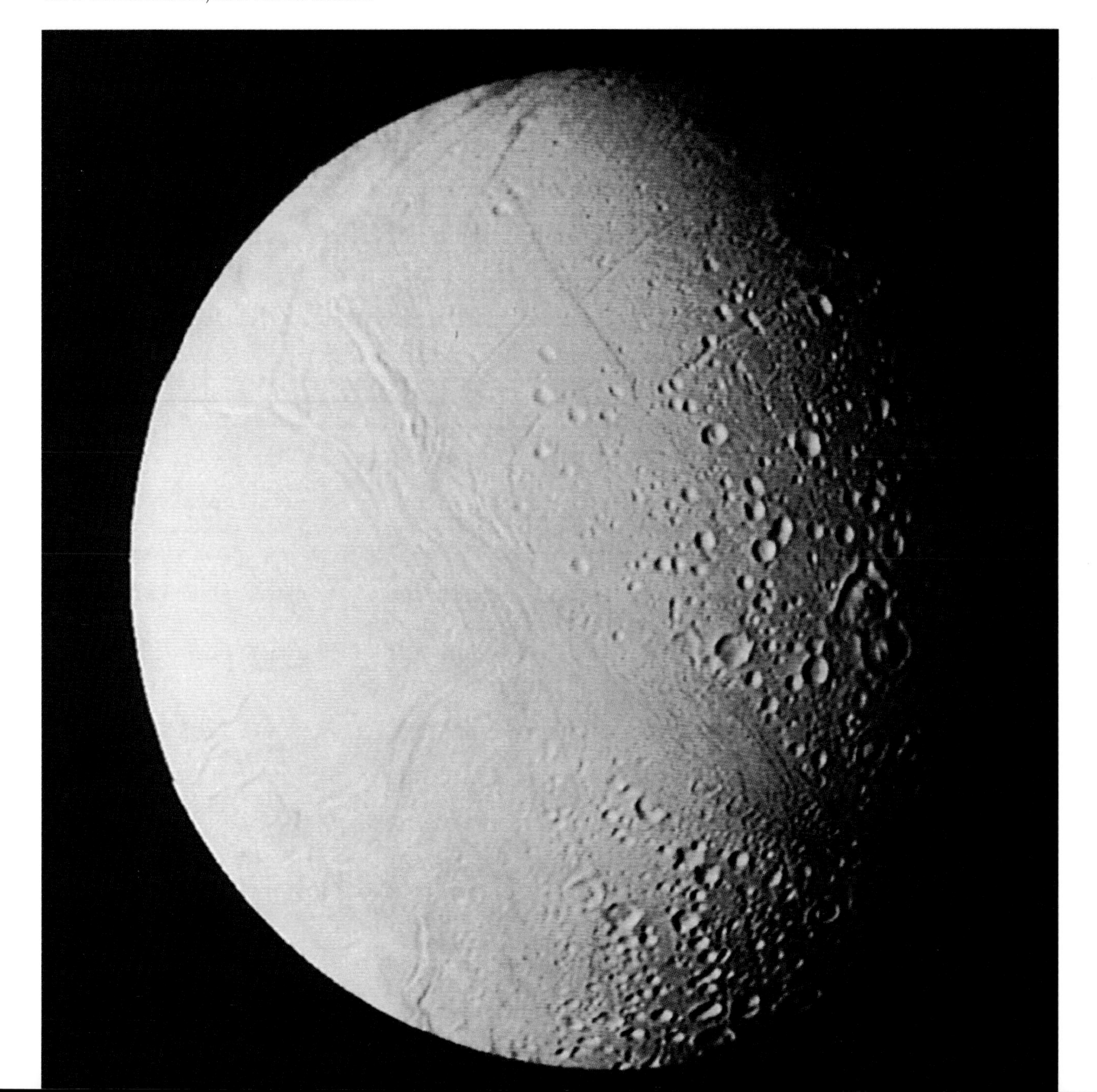

Above: Titan may reveal what Earth was like when life-forms first appeared. *(NASA)*

Studies by Hubble, passing planetary probes and radar scanning from Earth, make a strong case for finding natural tholin on Titan. If this is true, then deep beneath thick Titan atmosphere, primitive life-forms swarm through seas of methane. Radar indicates a mixed surface, including high ground in the form of at least one major continent. Hydrocarbon snow gathers in huge drifts on

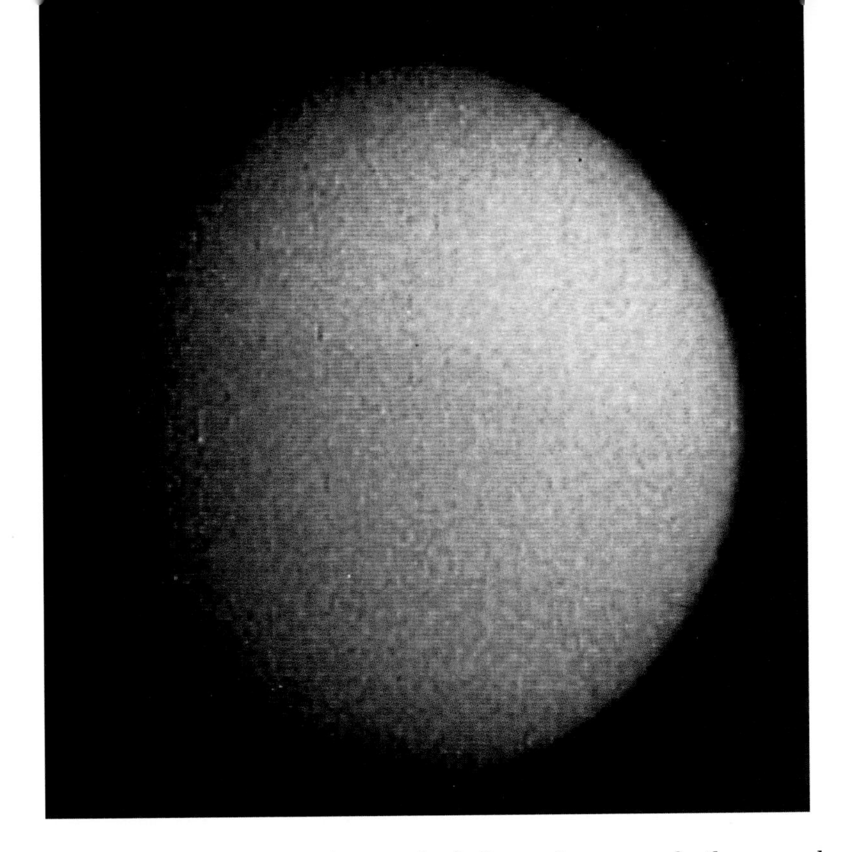

Left: Seen here in this Voyager photograph, the orange Titan is 3,200 miles in diameter. It is the only moon in the solar system with an atmosphere. *(NASA)*

that surface, often turning to slush from showers of ethane and methane.

Hidden even deeper within Titan are what only recently were discovered on our own planet—thermal vents several miles below our oceans, through which raw materials for life-forms pour forth in a steady torrent. Titan will become the target of elaborate probes dropping into the atmosphere, releasing instrumented balloons to float about the globe, and sending powerful instruments to the surface.

Titan hides its secrets in thick haze, but its parent world presents multicolored rings of asteroids, chunks of rock and ice, and trillions of smaller particles down to dust motes, reflecting the distant sun in stunning beauty. Thousands of major rings and ringlets swirl in a global dance about Saturn. Their shape is extraordinary, for they may be the thinnest surface of the entire solar system. If

Below: From left to right we see five of Saturn's moons. First, Tethys and its huge canyon system, followed by Dione, then the heavily cratered surface of Mimas, the enhancing moon that is Rhea, and completing the lineup, tiny Hyperion only 235 miles across. *(NASA)*

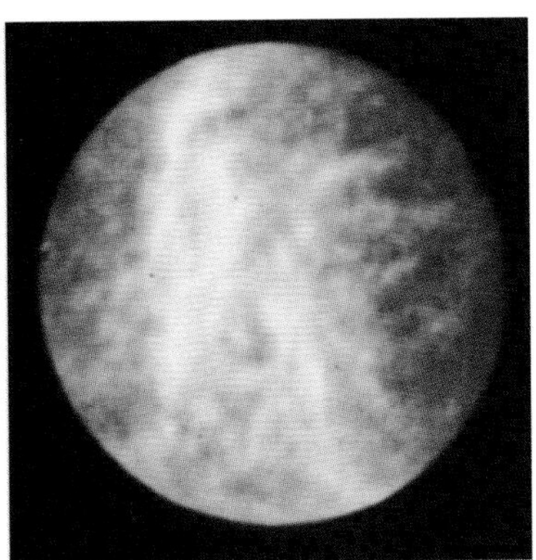

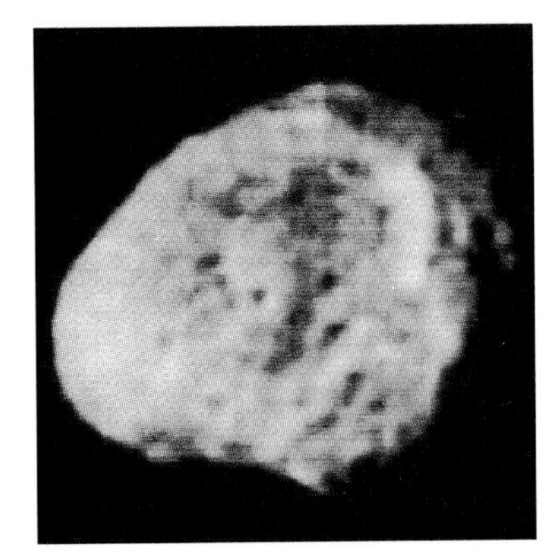

Above: Two of Saturn's moons, Tethys and Dione, are seen in this Voyager picture that shows clearly the beautiful planet's multicolored rings. *(McDonnell Douglas Aerospace)*

we made a model of Saturnian rings as a sheet of glass a hundred feet in diameter, that model would be thinner than the page you're now reading.

The rings dominate the planet from every aspect, gleaming with light and casting great shadows. Sometime in the early growth of the solar system, a planetoid passed dangerously close to Saturn. Great tidal forces of gravity tore apart the lifeless world. Chunks of rock and ice of all sizes fell into orbit about Saturn, swirling and tumbling as their own gravity sought to form them into a new moon of the gas giant.

But Saturn's domineering gravity prevented these chunks from massing into a solid world. Over many millions of years, the tidal forces separated the debris of a shattered planetoid into thousands of individual ringlets encircling Saturn.

When close-up photographs of Saturn reached Earth from the Voyager probes on their way to exit the solar system, another impossible physical display baffled astronomers. Rings and ringlets are held in gravitational suspension, a fine and equal division of the trillions of separate pieces. But the photos showed startling clumps, kinks, and braids in many of the rings. The mystery vanished when close examination of the pictures showed small moons moving through the plane of the rings and ringlets, their gravity tugging and shoving at the matter about them, as if invisible hands were braiding rings and ringlets of planet-sized strands of hair.

Below: The brilliant rings and ringlets of Saturn reflecting the light of the sun. *(NASA)*

Right: Photos from the Voyager probes revealed "small moons" moving through the plane of rings and ringlets. *(NASA)*

Below: Saturn's rings are the remains of a shattered planetoid that passed too close to Saturn and was torn apart by the force of its gravity. *(NASA)*

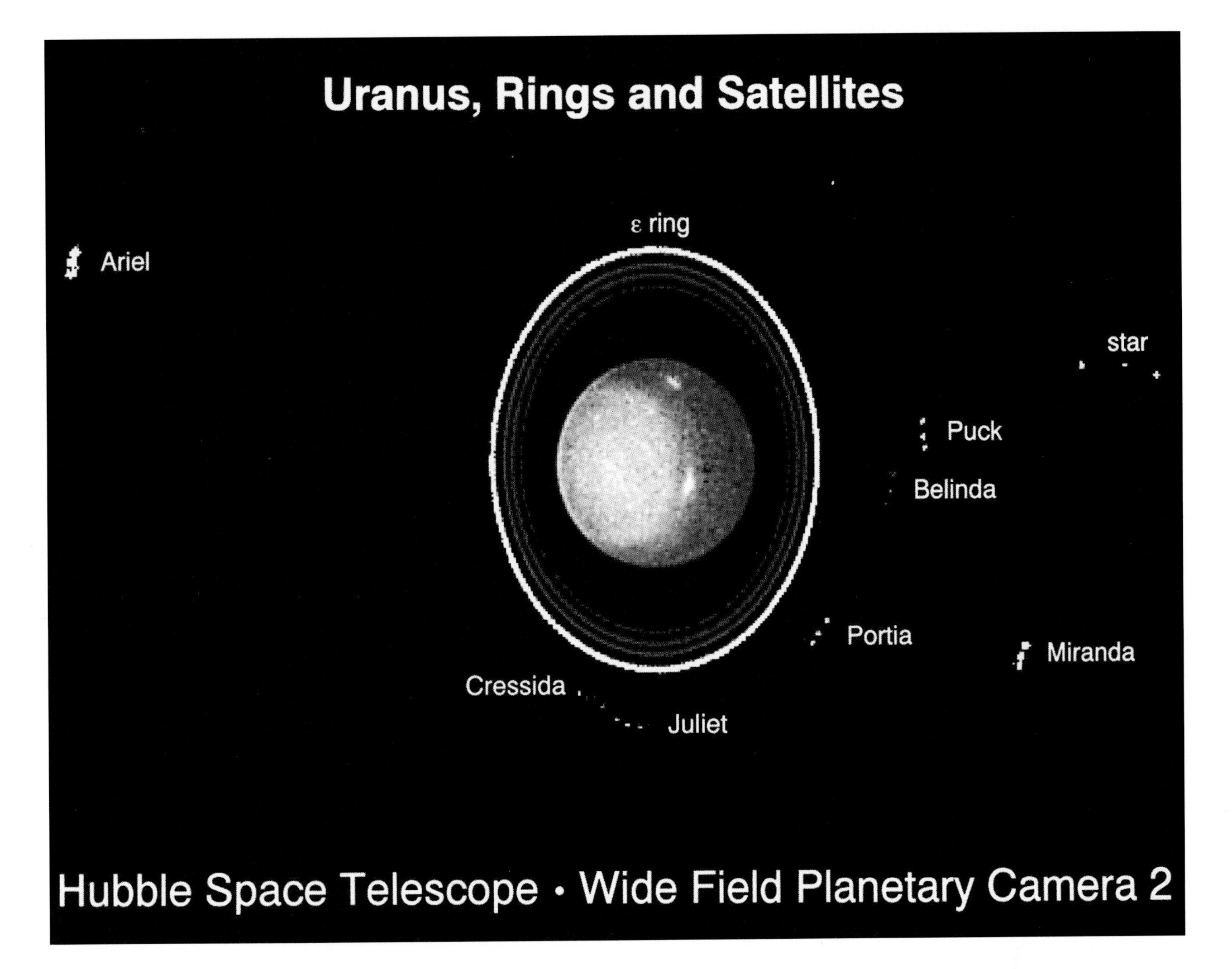

Above: On August 14, 1994, the Hubble Space Telescope aimed its cameras at Uranus, providing a fresh look at the planet's rarely seen rings, five of Uranus' inner moons, and bright clouds and a high altitude haze above the planet's south pole. *(Space Telescope Science Institute and NASA)*

URANUS

In the late summer of 1994, the Hubble Space Telescope cameras aimed at the seventh planet outward from the sun. At that moment Uranus was 1.7 billion miles distant from Hubble. Three new images of the blue-green planet were immediately acclaimed by NASA scientists as "spectacular." Great dark hoops girdled the mysterious orb that is more than four times the size of Earth. Hubble captured its extraordinary photos by looking at Uranus head-on and framed by its 11 concentric rings of dark dust.

Uranus posed for its portrait as no other world in the solar system, as the planet "that fell over on its side."

Most planets spin on an essentially vertical axis or one that is slightly tilted, like a toy gyroscope spinning on a point. Uranus glides about the sun like a gyroscope rolling along its rim.

Hubble performed its outstanding photography with the Wide Field Planetary Camera, revealing the dark rings and five of the inner moons in what elated NASA scientists cheered as a "giant celestial bull's-eye."

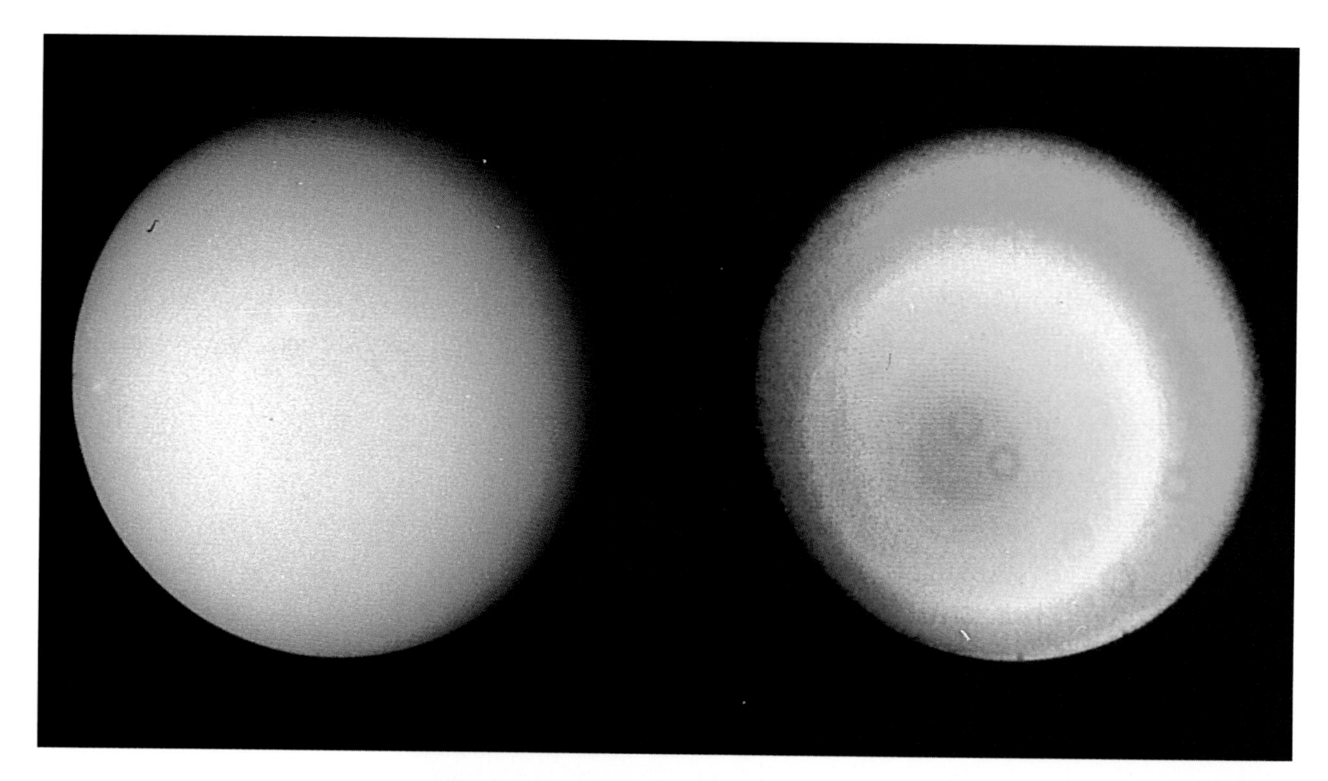

Above: Unlike other planets, Uranus spins on its side, its axis almost horizontal. This true-color shot of Uranus on the left and false-color view on the right were taken by Voyager 2 from a range of 5.7 million miles. *(NASA)*

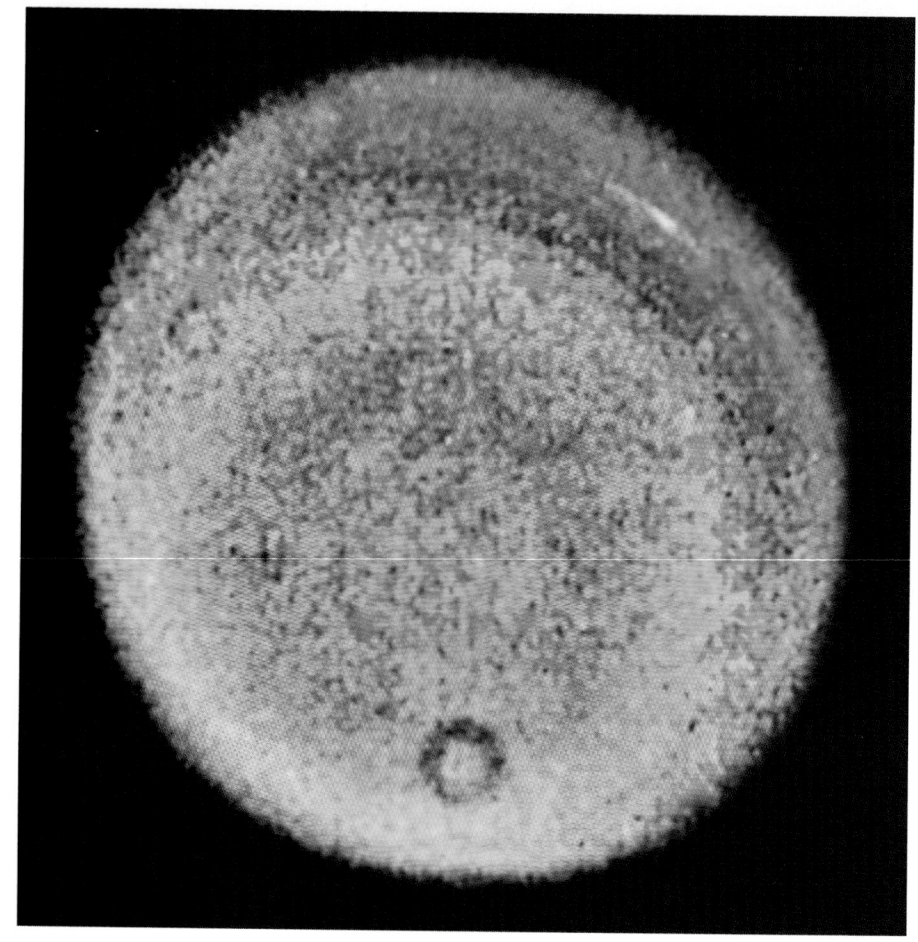

Right: Uranus receives only 1/400th of the sunlight that reaches Earth. This Voyager 2 photograph reveals bright clouds racing and a smoglike haze over the south polar regions. *(NASA)*

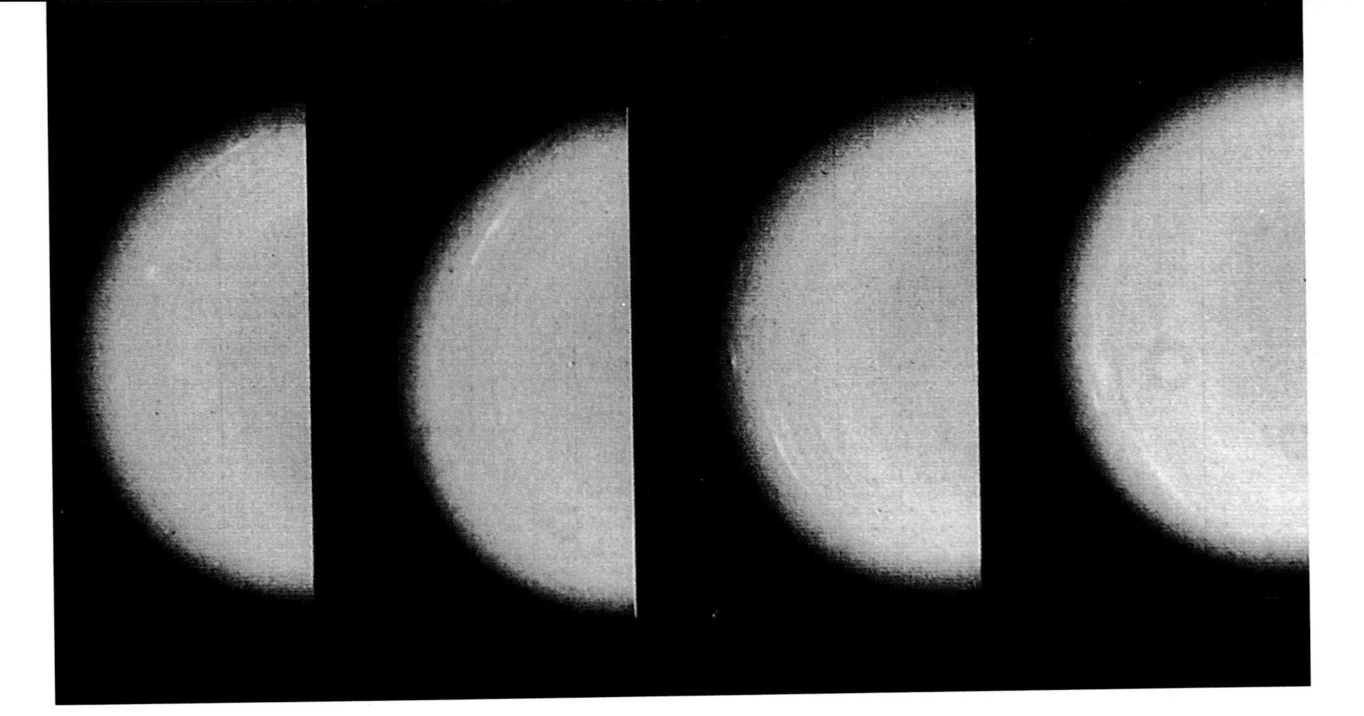

Uranus receives only one–four hundredth of the sunlight that bathes Earth. Photography from a distance of 1.7 billion miles of a world shrouded in eternal night had long been considered impossible. Hubble's camera reach revealed bright clouds racing at high speed over the world, as well as a smoglike haze hovering over the south polar regions.

Above: Hubble and Voyager have provided a new image of Uranus. These time-lapse images show cloud movements, indicating an atmosphere rife with huge storms. *(NASA)*

Until Voyager 2 and Hubble, Uranus was considered simply a great planet locked in bitter deep freeze, silent, forbidding, lacking even the atmospheric turbulence and inner heating of Jupiter and Saturn.

Voyager and Hubble brought forth a world that dumbfounded science. Uranus spun its multiple webs of ice, rock, and dust rings, as well as 15 moons in another miniature solar system. But what commanded revisions of all astronomy texts were the huge storms that sped across the planet, booming upward in Earth-like thunderstorms through the atmosphere of hydrogen, helium, methane, acetylene, and other hydrocarbons. Violent weather bands, similar to those of Saturn, girdled the planet, with Uranian wind speeds of nearly 400 miles per hour.

Above: Uranus is bathed in an unexplained radiance known as electroglow, seen in this Voyager 2 farewell shot of a crescent Uranus from a distance of 600,000 miles. *(NASA)*

The moons proved as surprising as the planet; most of them are frozen conglomerates of rock and water ice. The exception is Miranda, which is a misshapen crude ball once smashed into a storm of debris by asteroid impact. The flotsam of the shattered moon accreted rapidly, but the new gathering of material lacked heat. When violence subsided, Miranda was a moon unlike any other body in the solar system—mashed together unevenly, its surface cracked and pounded like a crudely packed frozen mud ball.

Despite eternal night, Uranus is bathed in an extraordinary radiance known as electroglow. A sphere with temperatures down to minus 350°F *should* be no brighter than the darkest night on Earth, but Uranus' atmosphere is agitated by a force still unknown to astronomers. Not only is the electroglow visible to Hubble's far-seeing cameras, but whatever strange forces roil within Uranus blow its atmosphere tens of thousands of miles into space, flowing as ghostly mist into the rings about the planet.

It is an extraordinary world that establishes a new and different yardstick with which Hubble will search for planets of other stars.

Below: Uranus has dark rings of ice, rock, and dust, as well as 15 moons. This picture taken by Voyager 2 shows a view of Uranus over the moon Miranda's horizon. Uranus is 65,000 miles away and its thin rings can clearly be seen over its blue-green shrouded surface. *(NASA)*

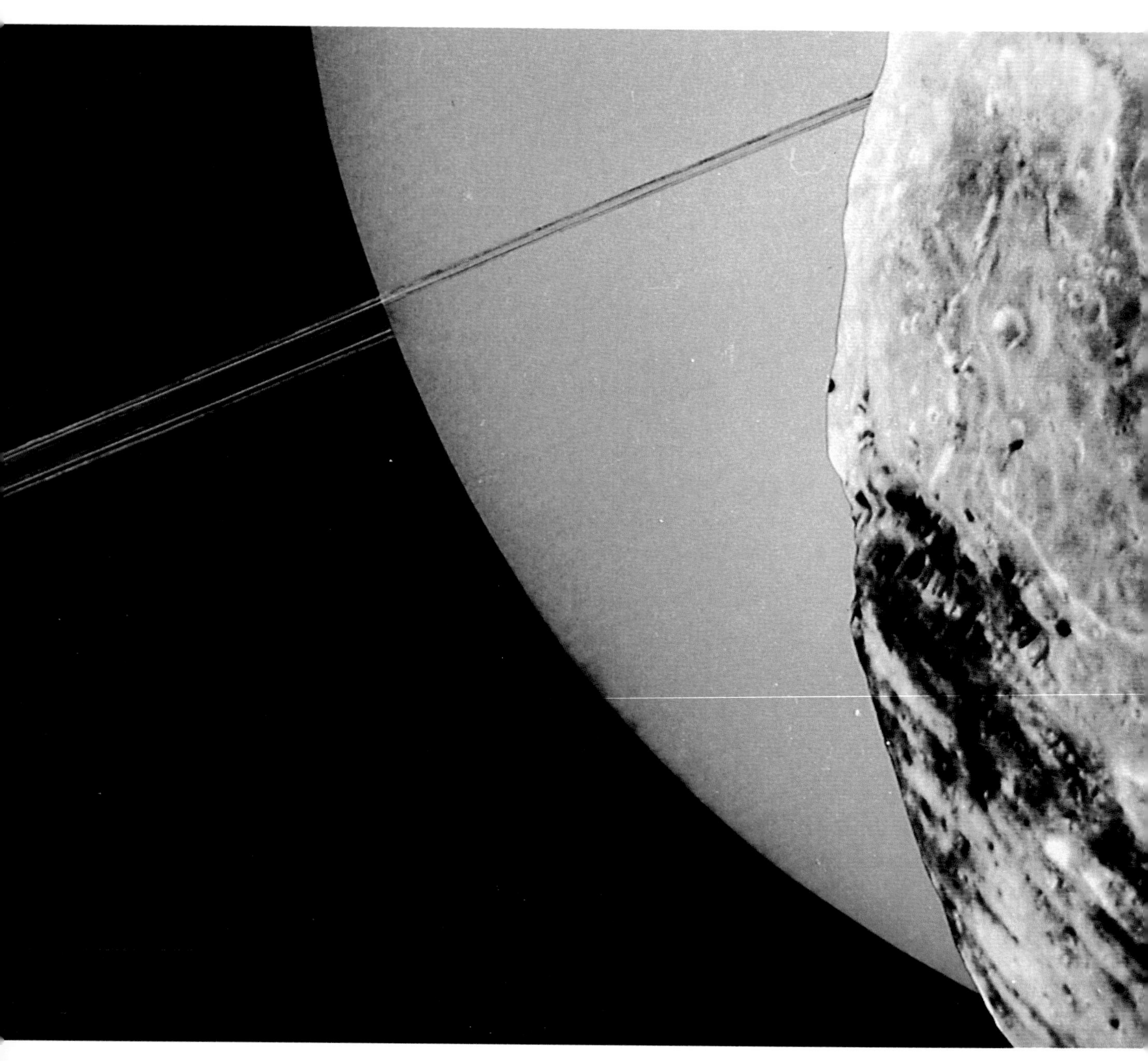

Above: The largest and fiercest storm on Neptune is seen in this Voyager photograph. It is called the Great Dark Spot, and its 1,500 mile-per-hour winds rotate counterclockwise. *(NASA)*

NEPTUNE

Uranus had always been judged a great ice-frozen sphere. Dead and unmoving, it was a sepulchral globe without the slightest stir for billions of years. This image went down the drain with the probing instruments of Voyager and Hubble.

Beyond Uranus lay Neptune, nearly three billion miles from Earth, four times the size of our world, locked in dark mystery deeper than Uranus. Finally eight moons were discovered circling the planet, and only 14 years ago astronomers detected what appear to be rings about the great frozen gas ball.

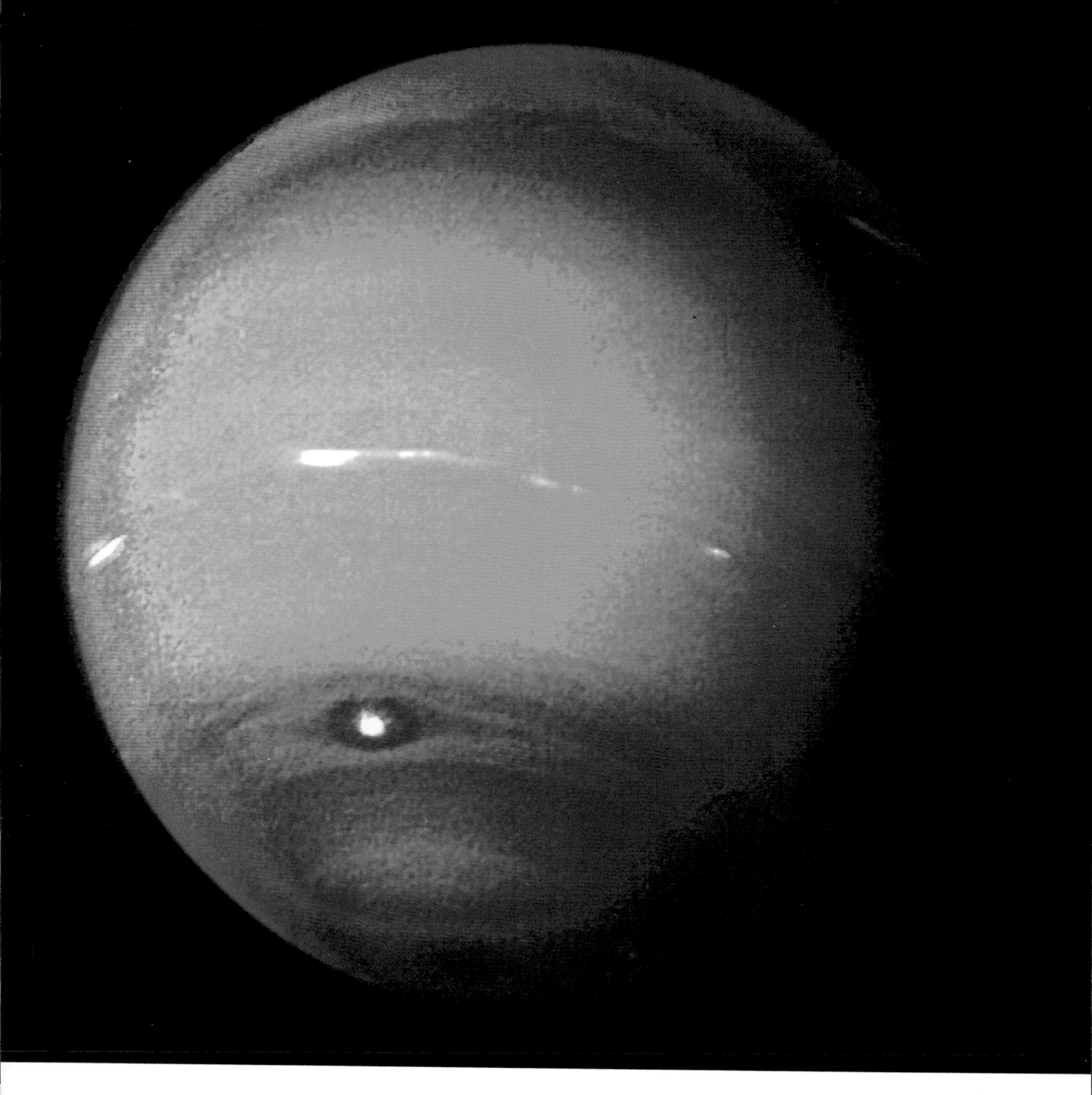

Above: As Voyager 2 made its approach, it snapped this color image of Neptune, the blue and aqua gas planet. Neptune's red, semitransparent haze can be seen covering the faraway world. *(NASA)*

These were the parameters with which science judged the first huge planets circling other stars far beyond our solar system. Huge, lifeless gas giants, suspended forever in a terrible deep-freeze.

Then the Voyager 2 space probe, racing away from Earth for 12 years at an average speed of 42,000 miles per hour, slipped past this last gas giant of our solar system. Instruments and cameras sent data on the long journey back to astronomers hoping for new information from a world so distant our sun was but a bright star in a black sky.

Left: Eight moons were discovered circling Neptune, and in this flyby picture taken by Voyager 2, Neptune and its largest moon, Triton, can be seen. Triton is the smaller crescent. *(NASA)*

Once again the astronomy textbooks were tossed into the trash heap. Scientists of NASA and Jet Propulsion Laboratory were excited, fascinated, but above all, confused.

"What they found," stated a NASA report of the astonishing revelations of Neptune, "will force scholars to rewrite the astronomy textbooks, and scientists to adjust their views of the solar system's other giant planets."

The faraway world always believed embalmed in rock-hard ice came alive. Some astronomers had predicted Neptune would

Below: This montage of six views, taken by Voyager 2 through various camera filters, gave altitude data on Neptune's clouds. *(NASA)*

in some ways mimic activity found on Uranus; they were the faithful few, while the larger body of scientists believed Neptune was dead and buried in its permanent icy vault.

The few believers triumphed in more ways than anyone believed possible. Voyager 2 was sped so fast past Neptune that it could have crossed the entire United States in four minutes flat, hardly enough time for a grandiose picture show from near the end of the solar system.

The photographs stunned the Voyager team. A vibrant, active, even turbulent great blue planet appeared before them. Even at this distance of billions of miles, ultraviolet radiation from the sun destroyed methane high in Neptune's thin atmosphere, creating complex hydrocarbons. Particles freed by radiation sank deep into the thick atmosphere, froze into ice, and were then *warmed* within the lower levels of the planet. From a world supposedly ice-dead, hydrocarbons evaporated and surged upward to create huge storms sweeping about the globe.

Above: Fourteen years ago, astronomers detected what appeared to be rings girdling Neptune. They were right. Voyager 2 took these pictures of the great frozen gas planet's rings. Neptune's ring system is shown here in the left picture in two exposures lasting nearly 10 minutes each. Details of Neptune's rings can be seen in the picture on the right. *(NASA)*

Neptune was *dynamic*. Oval-shaped storms waxed and waned with astonishing power and turbulence. High over the raging storms, especially the largest, named the Great Dark Spot, sped wispy white clouds, eerily like cirrus and lenticular clouds of the dynamic, heat-driven weather engine of Earth. Other storm systems circled the 30,775-mile-diameter world in only 16 hours.

Storms shredded the atmosphere with the fiercest winds of any planet in all the solar system, blowing east to west at 1,500 miles per hour—winds so powerful they were like the shock waves of atomic bomb blasts in our atmosphere. By comparison, a hurricane with 200-mile-per-hour winds is the mightiest storm ever to rage on Earth.

Magnetic and electrical fields swirl wildly about Neptune, creating astonishing auroras (much like the northern and southern lights of Earth) so energetic they race across and over most of the planet.

But one of Neptune's eight known moons stole the show from

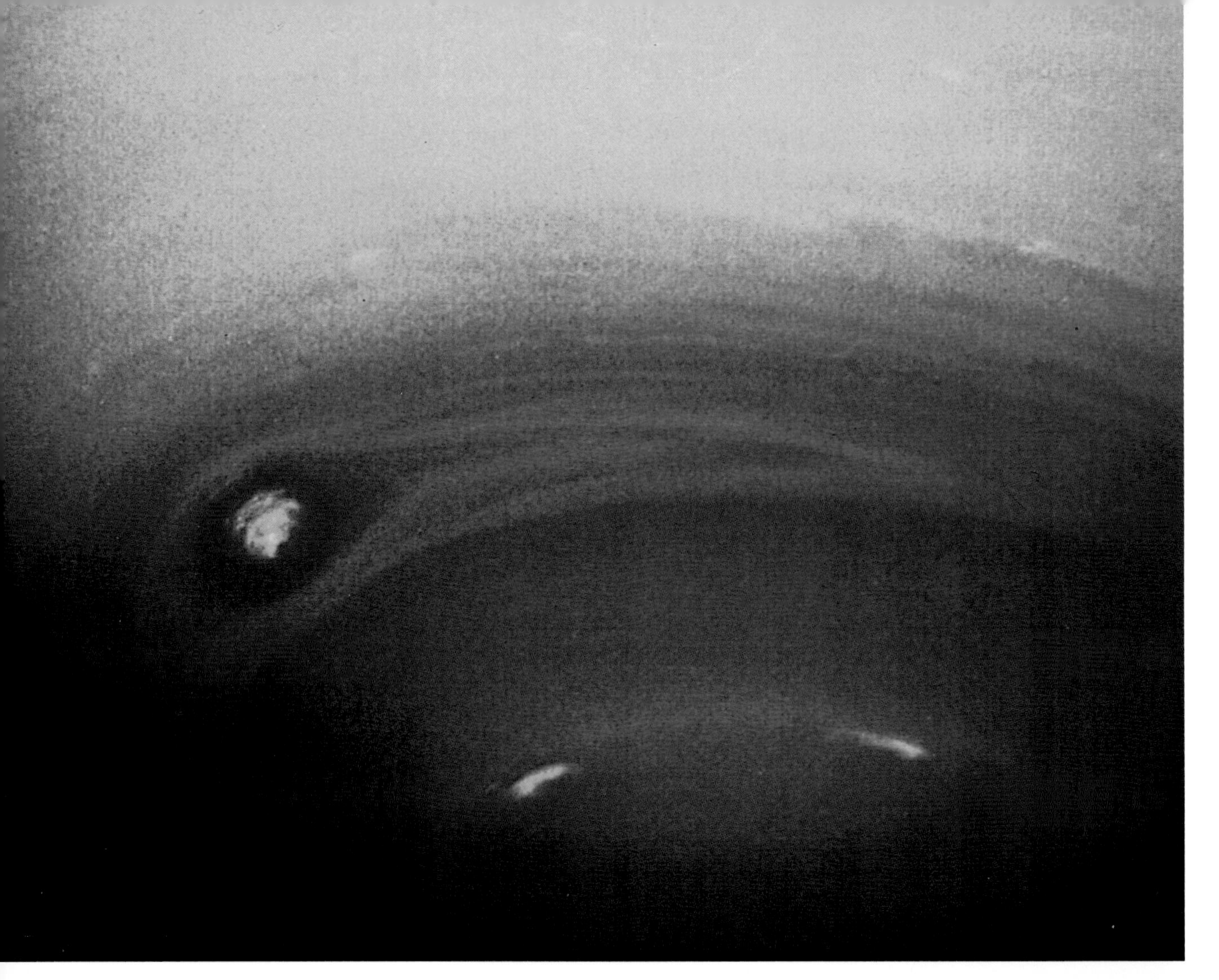

Above: A Voyager photograph of the cloud systems in Neptune's southern hemisphere. *(NASA)*

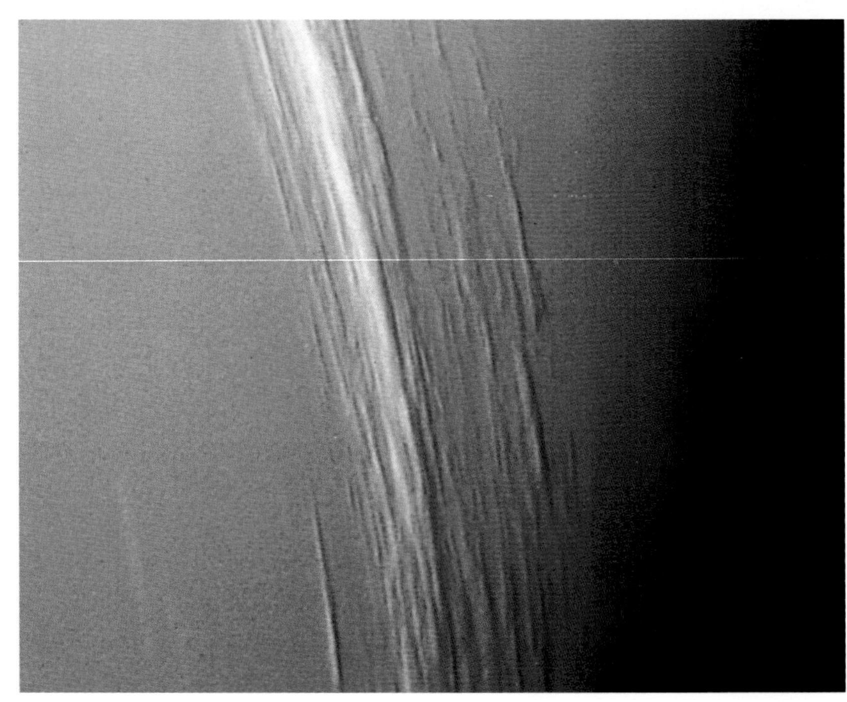

Right: Storms like these high-altitude cloud streaks shown in this Voyager picture shred Neptune's atmosphere with the fiercest winds of any planet in all the solar system, with wind speeds at 1,500 miles per hour. *(NASA)*

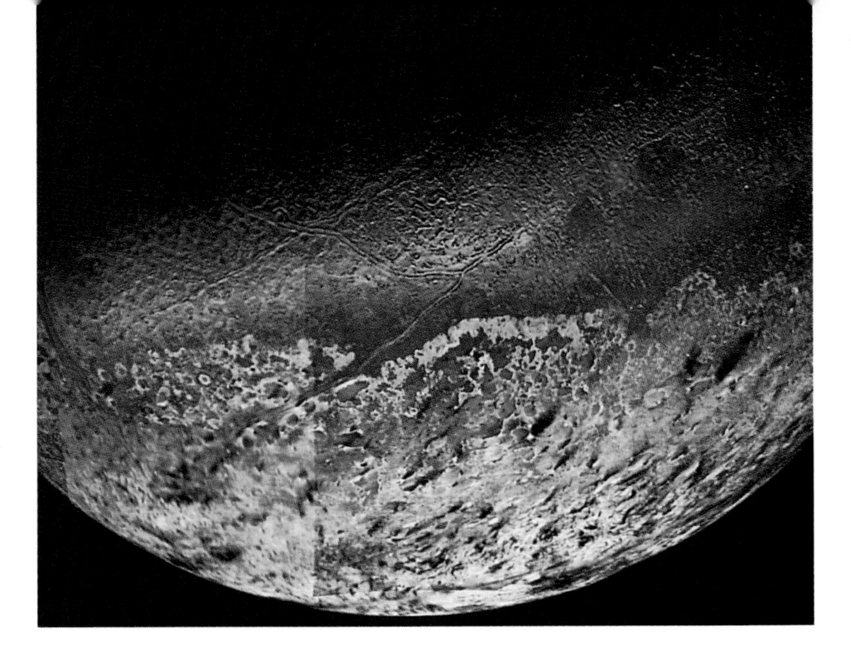

Left: A Voyager photograph of Triton's southern hemisphere. *(NASA)*

its host world. Astronomers were astonished to find that Triton, only three-quarters the size of our moon, was also a dynamic globe of totally unexpected activity. The temperature of Triton is the coldest ever measured of any body in space—391°F below zero. With this all-embracing cold it simply didn't seem possible the Neptune moon could be anything but frozen ice and rock.

The supposed dead world again tumbled the textbooks into oblivion. Astronomers gaped at pictures of eruptions spewing nitrogen gas and clouds of charcoal-black dust miles into space. What seemed impossible had been captured, unquestioned, on the amazing pictures sent back to Earth. But how could a cold-storage moon be spouting geysers?

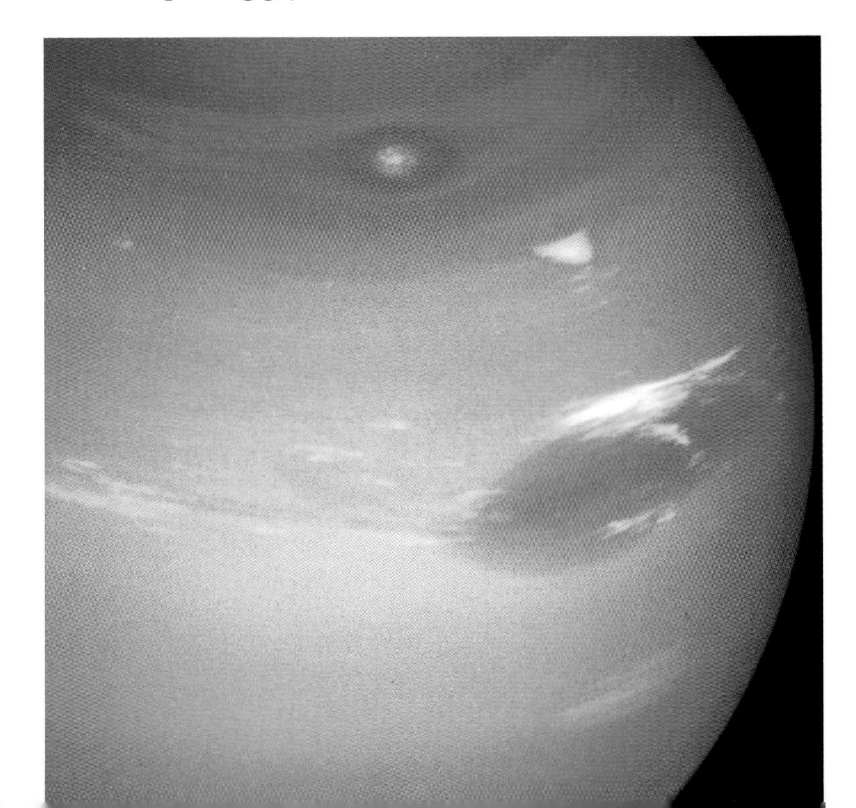

Left: Voyager's astronomers found Neptune's atmosphere to be riven with violent storms of extraordinary magnitude. *(NASA)*

Something astonishing swept along Triton's surface—*ice volcanism.* Could there really be volcanic-like eruptions on a world with temperatures so close to absolute zero? The impossible was happening, and equally astounding to scientists was that most geologic structures on Triton had been formed from frozen *water*—water as rigid as granite transforming from slush ice to gas and then bursting violently upward for miles. Gas and dust swirling in the winds awaiting their arrival then blew far downwind from the icy volcanic vents.

Lavalike flows move sluggishly across the surface. It seemed as if Triton were alive, this amazing moon where glaciers and rocks formed from methane and nitrogen flow like molten lava on Earth. High above, wind streaks and fleecy clouds race across the surface, carrying all the way to another unexpected marvel—polar caps on an already "impossible" moon.

Inner heat could cause some of this activity, but never enough to cause geysers erupting across the entire planet surface. The sun, from so far away, lacks the warmth to soften ice as rigid as

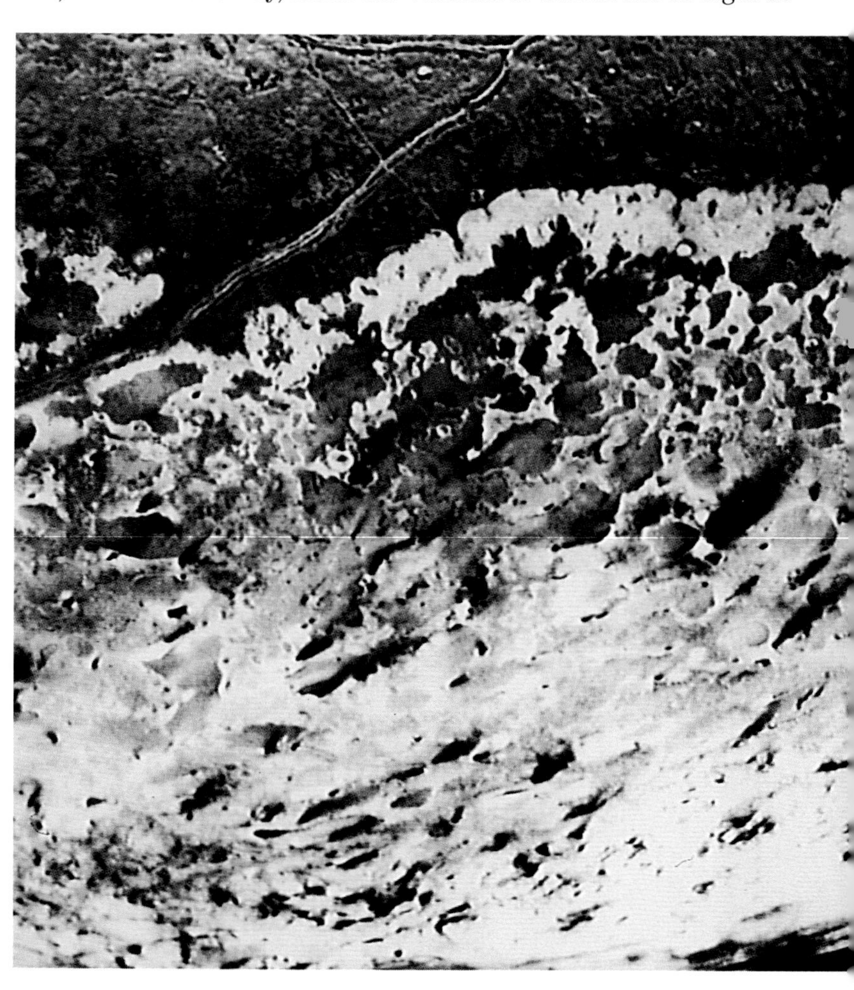

Right: About 50 dark plumes mark what may be ice volcanoes in this Voyager picture of Triton's south polar terrain. *(NASA)*

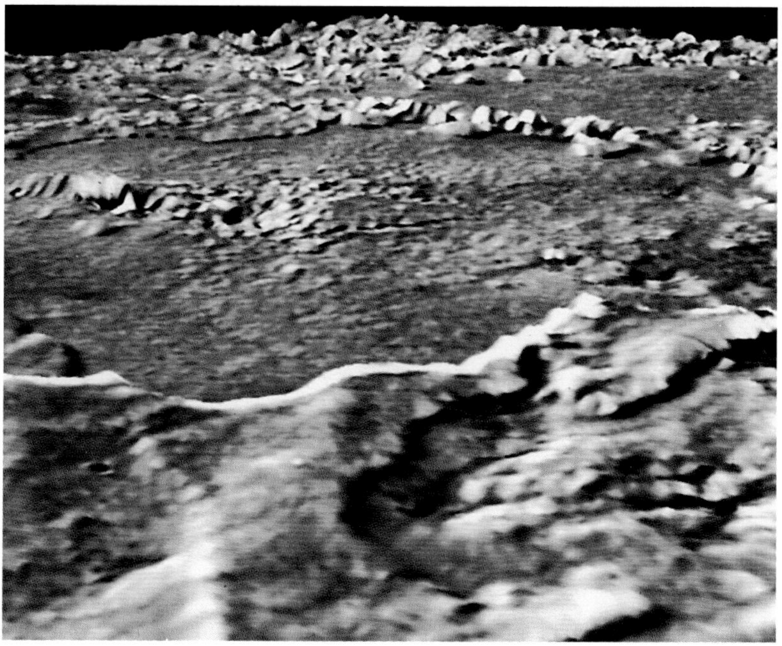

Above: Depressions, as seen in this picture taken by Voyager, may be caused by melting and collapsing of Triton's icy surface. *(NASA)*

Left: Lavalike flows are seen in this Voyager picture of the surface of Triton. *(NASA)*

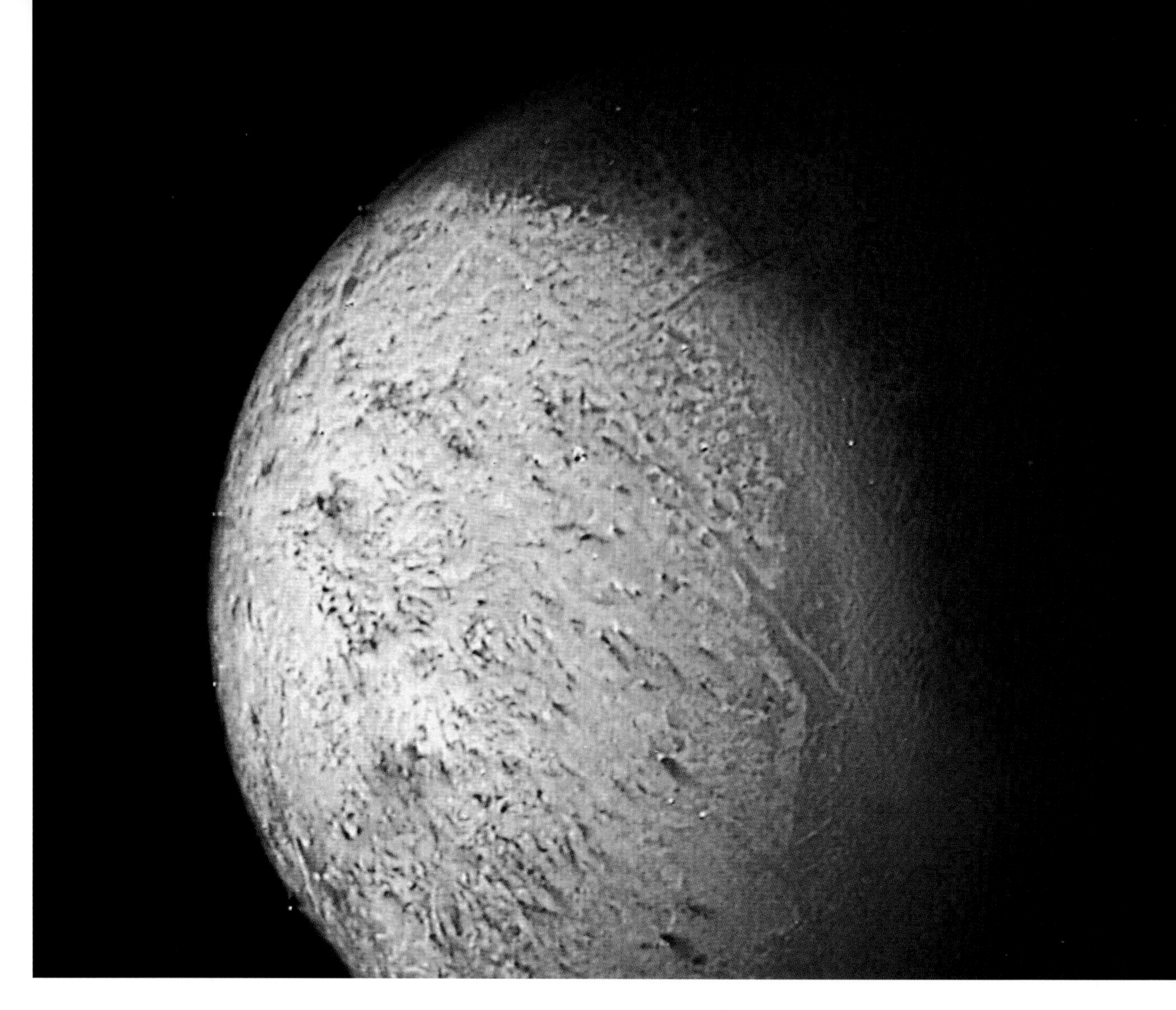

Above: Another Voyager photograph of Triton. *(NASA)*

granite. The solution lay in an extremely thin atmosphere diffusing to 500 miles above the globe.

Triton undergoes a heating cycle measured in millions of years. The sources are the stars of the galaxy, and the ejecta of exploding suns millions of light-years distant. Cosmic rays wash against and deep into Triton from space. The radiation transforms methane into black organic powder, which again undergoes a change into thicker and crusty material.

Then, in a strange departure from the hellhouse of Venus, nitrogen ice sliding over the crust reacts to even the pale and weak radiations from the sun. The ice traps heat beneath its layers. Finally, the heat expands the dust and gases beneath the ice belt above—and explodes outward in the geysers that form a frozen, dark Yellowstone National Park near the edge of the solar system.

To the scientific teams managing the Hubble Space Telescope, all the rules of planetary study have been shoved aside to accommodate the reality of the frigid yet volatile gas giants of

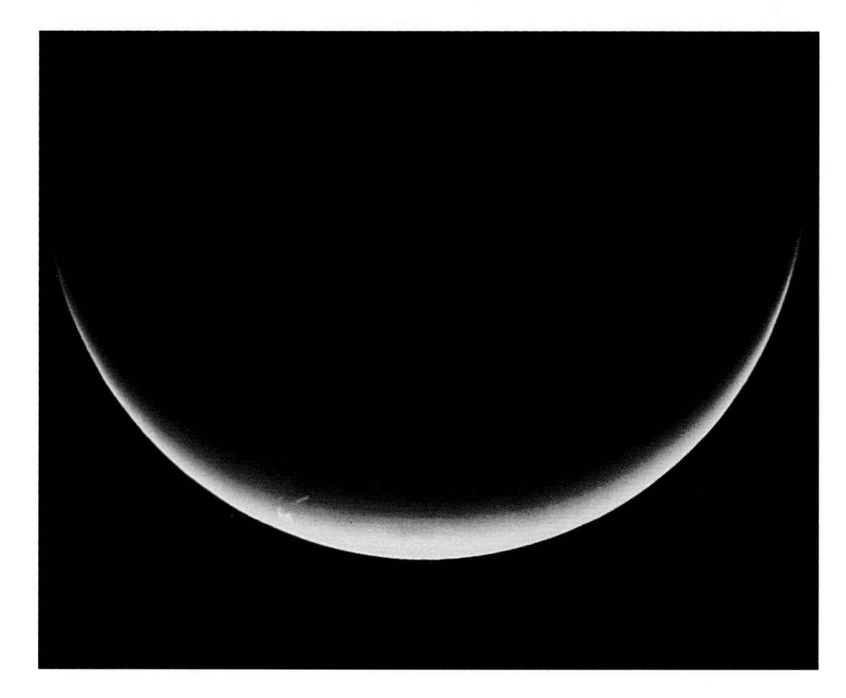

Left: As Voyager 2 left the solar system, it took another look in this final photograph of Neptune's limb. *(NASA)*

our most distant solar system worlds. Now that the measuring yardsticks have been created with accuracy, the Hubble teams know they will most likely encounter similar planets about other stars. The parameters are no longer shrouded in darkness and ignorance.

Knowing where to look, and what to look for, endows Hubble with ever-greater possibilities of finding not only other gas giants, but other smaller worlds that may direct us to the greatest wonders of all—planets like Earth.

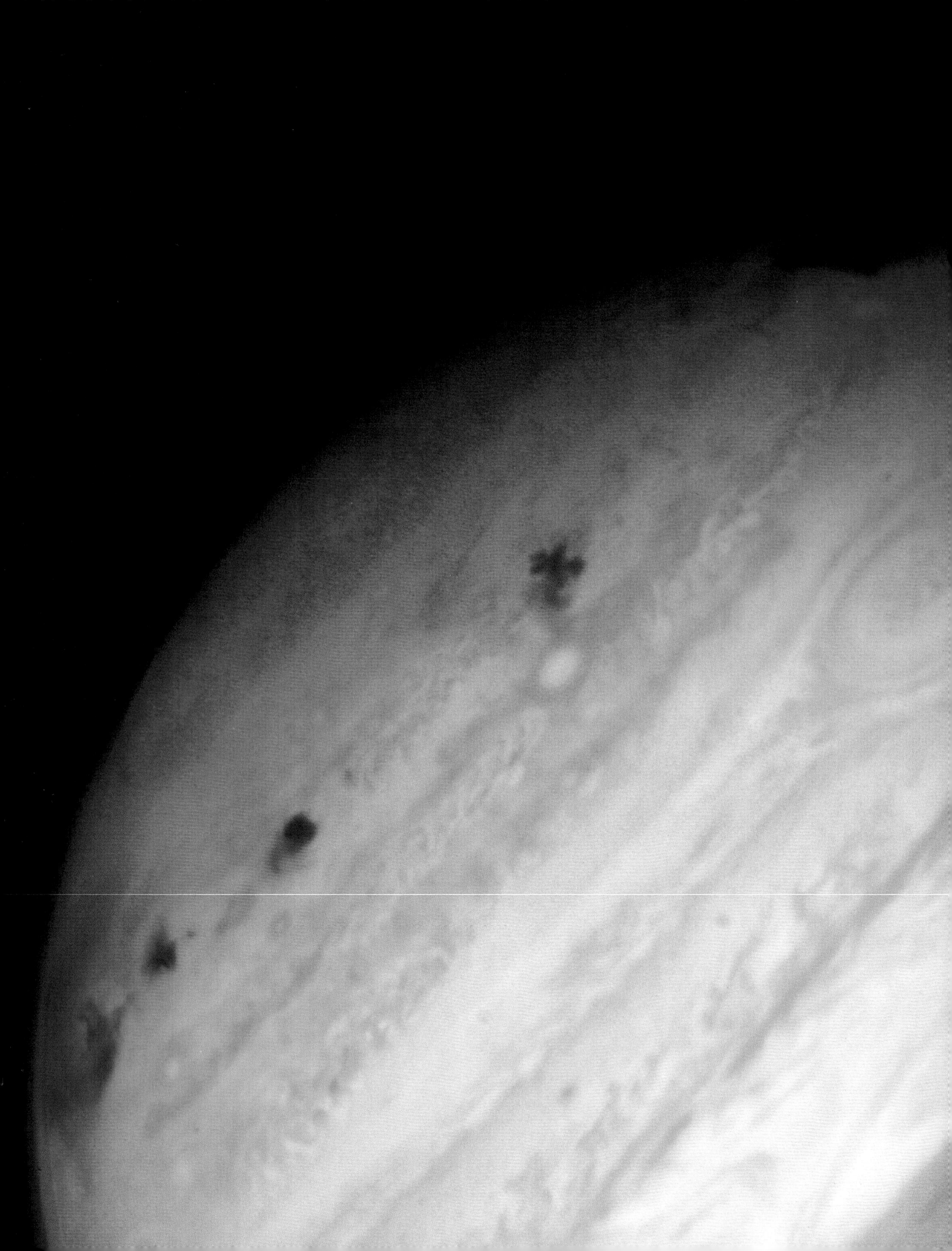

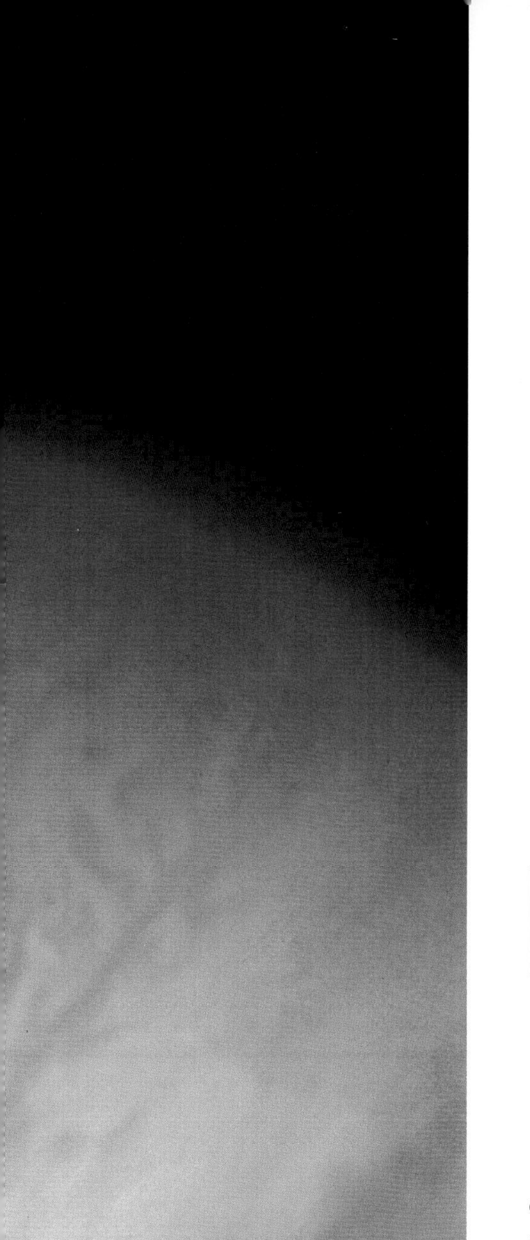

9 PLUTO, OORT, AND NEMESIS

The edge of our solar system is forbidding, icy darkness. Nearly 3.7 billion miles from the sun, the last known major world rolls silently along the outer rim of our family of planets, a dark, small sphere 1,430 miles in diameter. Once every 248 years—a greater time span than the entire history of the United States—Pluto completes a single orbit around the sun, dragging along its single known moon, Charon. They are a remarkable pair, the closest any two objects come to being measured as a double-planet system.

Until the advent of Hubble and its powerful cameras, Pluto remained an unfocused, fuzzy image to astronomers. And now, once again, the textbooks are being rewritten. The surface temperature of this distant planet plummets to 40°F above absolute zero (minus 451°F). At this frigid existence it seems nothing could exist save unchanging, rock-hard ice, yet Hubble detected a thin methane atmosphere, a strange yellowish haze that hints at activity on the rocky world. Initial photographic and instrument scans were even more startling when they displayed polar ice caps distinct from the rest of the rigid surface.

Pluto is also a maverick among better-behaved orbiting

Previous page: Hubble took this picture of Jupiter, with its surface scarred by eight impact sites by pieces of Shoemaker-Levy. On the far right limb of the planet, extended haze at the edge of the giant planet can be seen over the bigger-than-earth, doughnut impact. (*Space Telescope Science Institute and NASA)*

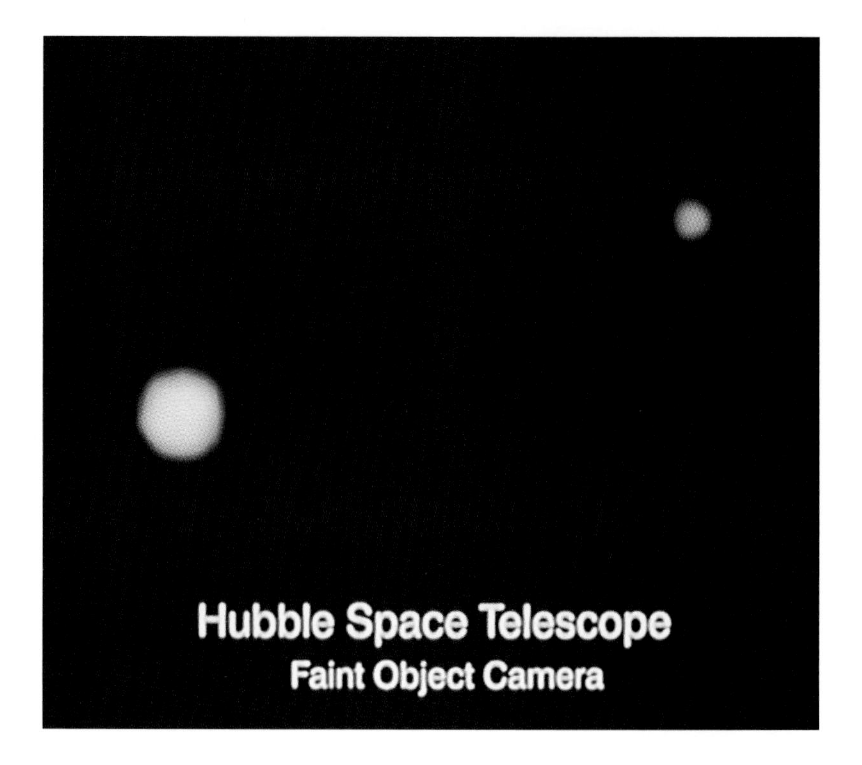

Right: On February 21, 1994, when Pluto was 2.6 billion miles from Earth, or nearly 30 times the distance between Earth and the sun, Hubble took this picture—the clearest view yet of Pluto and its moon. *(Space Telescope Science Institute and NASA)*

worlds. Unlike the eight other planets, Pluto periodically slides inside the orbit of Neptune, so that the two planets exchange honors for being the solar system's outermost globe.

But Pluto and Neptune may not be the outermost worlds. Some astronomers believe there is yet another planet far beyond the orbit of Pluto—a planet at least five times as dense as Earth, traveling in a severely elliptical orbit.

World astronomers began the search for the suspected planet in the 1980s. NASA scientists speculate that a large dark-star-type body, perhaps as huge as the sun, may drift in an orbit as much as 50 billion miles beyond the farthest sweep of Pluto and Neptune.

Yet this seems much too extreme for a member of the solar system still governed by the gravitational forces of our sun and planets. Nevertheless, deep-space observations by ground telescopes, as well as by Hubble, indicate that a distance of 50 billion miles or more is actually common for burned-out suns as companions of visible and energetic stars.

AGITATING THE OORT

Finding planets or clusters of rocky debris is always a goal for dedicated astronomers. Anything that affects comets orbiting the sun at *any* distance is cause for concern. Dislodged from ancient orbits, they are booted by collisions of gravity effects into new paths that send them sunward.

These are the times when the planets and moons become a cosmic shooting gallery. The Shoemaker-Levy comet traveled through the outer reaches of the solar system before being shoved by passing rocky bodies from its ancient orbital path to begin accelerating because of the sun's gravity. The comet—an ice-shrouded rocky mass—swept close by Jupiter in the distant past. Jovian gravity broke up the single comet nucleus and also tightened its orbital path, so that after another swing through the solar system, collision with Jupiter in 1994 became inevitable. From Earth observatories, Hubble, and the planetary probe Galileo, we watched a barrage of deadly debris slamming like shots from a giant repeating rifle into Jupiter. Change comet Shoemaker-Levy's target from Jupiter to Earth, and no one would be reading these pages today.

Below: The Shoemaker-Levy comet passed near Jupiter, the gravity of which broke up the single comet nucleus. Seen here, in this composite photograph by Hubble, is the comet prior to its collision with Jupiter. *(Space Telescope Science Institute and NASA)*

But from exactly where did Shoemaker-Levy begin its inward journey? Or other comets, like Swift-Tuttle, which slipped past Earth in November 1992 and will return in August of 2126? Or Arend-Roland, which whizzed by our world in 1957? In February 1976, comet West was a preview of Shoemaker-Levy. The central core of West blew apart under gravity pressure of the sun and several planets. Finally it became a string of at least four large rocky bodies, all of which, fortunately, swung around the inner solar system and began their return to the dim reaches beyond Pluto.

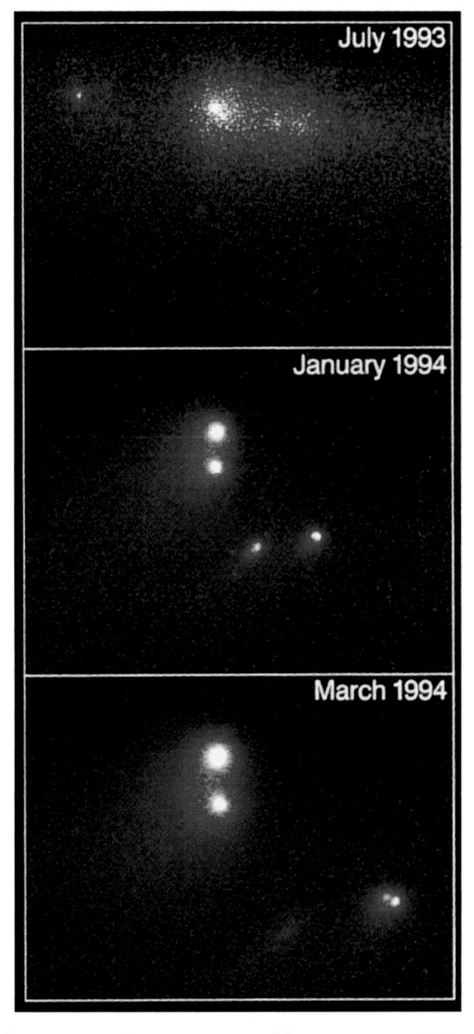

For many years, the ancients regarded the Star of Bethlehem—a violent flaring of light in the heavens—not merely as a theological event, but as the dazzling flash of a supernova, a great star that exploded in a blast brighter than a million suns. Chinese astronomers recorded the stellar flare-up nearly 2,000 years ago. Updated studies now lean toward the Bethlehem event as a comet that burst into visible light as a sudden blaze in the sky with an enormous tail stretching tens of millions of miles, sweeping much of the world with religious fervor.

Some comets leave a signature in violence, as did the Great Tunguska Event of June 30, 1908. A comet set ablaze by friction with our atmosphere sent a huge fireball roaring across central Siberia, described by terrified witnesses as "sheets of flame that split the sky in two." The comet lacked sufficient mass to reach the ground; it exploded in midair. Tens of thousands of trees burst instantly into flames, trains a hundred miles away were blown off their tracks, thousands of square miles were devastated, and the shock wave from the explosion swept twice around the world.

The Tunguska comet was no more than 100 yards in diameter, yet weighed more than a million tons.

Few space visitors match the regularity, brightness, or renown of Halley's Comet. In 1507 B.C., Chinese astronomers first recorded the brilliant glowing nucleus of the comet and its gleam-

Above: Two Japanese astronomers discovered the comet Ikeya-Seki in 1965. It soon became visible in the night sky, with a glowing tail 30 million miles long. *(U.S. Naval Observatory)*

ing tail stretching through space for millions of miles. What they saw was the ice-shrouded core and the fabulous plume of dust and gas blown away from the comet and illuminated by solar pressure.

Halley returned in A.D. 1066, exciting and frightening people throughout Western Europe. In 1301, it blazed through the skies across Italy, so bright it was seen clearly in daylight. Supernatural powers were attributed to this cosmic wanderer. The Great Comet of 1466 was actually Halley revisiting, but millions of people panicked when word spread that the huge glowing sphere and trailing plume was coming punishment from God, who would slaughter Christians for failing to win their war against infidels.

Not all comets that plunge into the inner solar system incite such emotional fervor. In 1965, two Japanese astronomers discovered a comet, then gave it their names—Ikeya-Seki. It soon became visible in the night sky with a glowing tail 30 million miles long.

Eighteen years later another flashed in from deep space on a death dive. Named Iras-Araki-Alcock, the comet made its appearance on May 8, 1983, as it bore steadily toward the sun, accelerating as it passed through the orbits of Mars, Earth, Venus, and Mercury. That proved its undoing. Subjected to the powerful magnetic fields and immense radiation of the four planets closest to the sun, it was vaporized, in a final flash of fire, to gas and dust.

Almost all comets originate from a vast ring of debris orbiting the sun far beyond Pluto. This is the Oort Cloud, so distant that the sun appears only as one among other bright stars in the black sky. The cloud is a zone of trillions of chunks of rock, dust, and ice, most of the particles small, others weighing millions of tons.

Any disturbance of the gravitational balance of this celestial debris can trigger a mass of this material inward toward the sun. The disturbance could be caused by a wandering asteroid or a

Below: Most comets originate from the Oort Cloud, a vast ring of debris orbiting the sun far beyond Pluto.

As seen streaking across the Milky Way in this U.S. Naval Observatory photograph, the Echo and other satellites are oftentimes mistaken for comets or meteorites. *(U.S. Naval Observatory)*

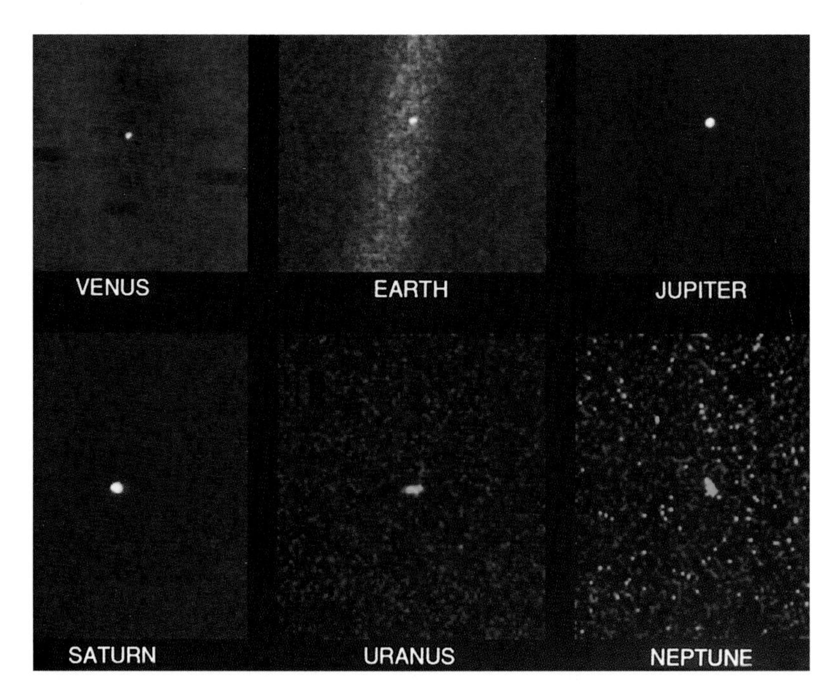

Right: This is a remarkable assembly of six of our planets as they might be seen by a visitor approaching us from beyond our solar system. In this montage one sees, from top left to lower right, Venus, Earth, Jupiter, Saturn, Uranus, and Neptune.

Voyager 1 took this first ever "portrait" of our Earth appearing as any other star from 4 billion miles in space. *(NASA)*

larger planetoid. It might be the effect of the still hypothetical Planet X, which is believed by many astronomers to exist, but cannot yet be confirmed by direct observation. But *something* obviously upsets the orbiting balance of the Oort Cloud to send the flurry of ice and rocks sunward.

In 1992, the science journal *New Worlds Explorer* reported two "trans-Plutonium objects" orbiting the sun at a distance of 4.4 billion miles. The newly discovered planetoids are 120 and 140 miles in diameter, are distinct from comet material, and follow regular planetlike orbits about the sun. They receive less than one–two thousandth the light lavished on Earth. Both planetoids show a reddish hue, and instruments measured their surface temperatures as only 30°F above absolute zero. Their positions and stable orbits lead astronomers to conclude that they do not disturb the Oort Cloud sufficiently to unleash comets toward the inner solar system.

Planet X—the theorized tenth planet—remains an unconfirmed source of comet dispersion from the Oort Cloud. Dr. John Anderson of the Jet Propulsion Laboratory believes there may be a dark stellar companion to our sun of the same size and mass, but in orbit 50 billion miles beyond Pluto.

And a final contender for a shaker of worlds is postulated by some NASA astronomers as a black hole, the collapsed remains of a massive star that orbits our solar system 100 billion miles beyond

Neptune. Impossible to see in visible light, its gravity is so powerful that anything sucked into its overwhelming density—even light—cannot escape its grasp. It orbits invisibly, known only by its distant gravitational effect upon our solar system.

Astrophysicist Richard Muller of the University of California at Berkeley, and associates Marc Davis and Piet Hut of the Institute for Advanced Study, stand by their theory that this Death Star has been the cause of mass devastation on all the planets. This Death Star, they propose, is that dark companion of our sun, and every 28 million years its orbit brings it through the Oort Cloud to wreak havoc among the mass of material, kicking loose *billions* of particles, small and huge. This research team points to the past mass disasters on Earth from these swarms of comets, noting that our planet sustained pulverizing blows 125 million, 92 million, 65 million, 38 million, and 12 million years ago. This regularity can only be, they theorize, the effects of the Death Star—Nemesis.

They may be in error about the Death Star; Planet X, other scientists believe, may be the culprit to endanger future life on Earth.

That's what space probes, and now Hubble, are for. Hubble will continue to extend its reach to capture light from all areas of the universe, and part of that search will be to unfold the mysteries from the edge of our own solar system.

Other probes will follow where Pioneers and Voyagers, now already moving beyond our solar system toward distant stars, have led. As they depart on their long, lonely journeys, they bid farewell to our sun and will send more pictures like those returned from Voyager 1, a remarkable gathering of worlds as they might be seen by a visitor approaching us from beyond our solar system.

Our worlds will appear, individually or grouped, as they did in a one-of-a-kind montage of planets by Voyager 1—the only photograph ever made of Venus, Earth, Jupiter, Saturn, Uranus and Neptune from outside of our solar system.

Now begins our journey outward, not only through space but back through time.

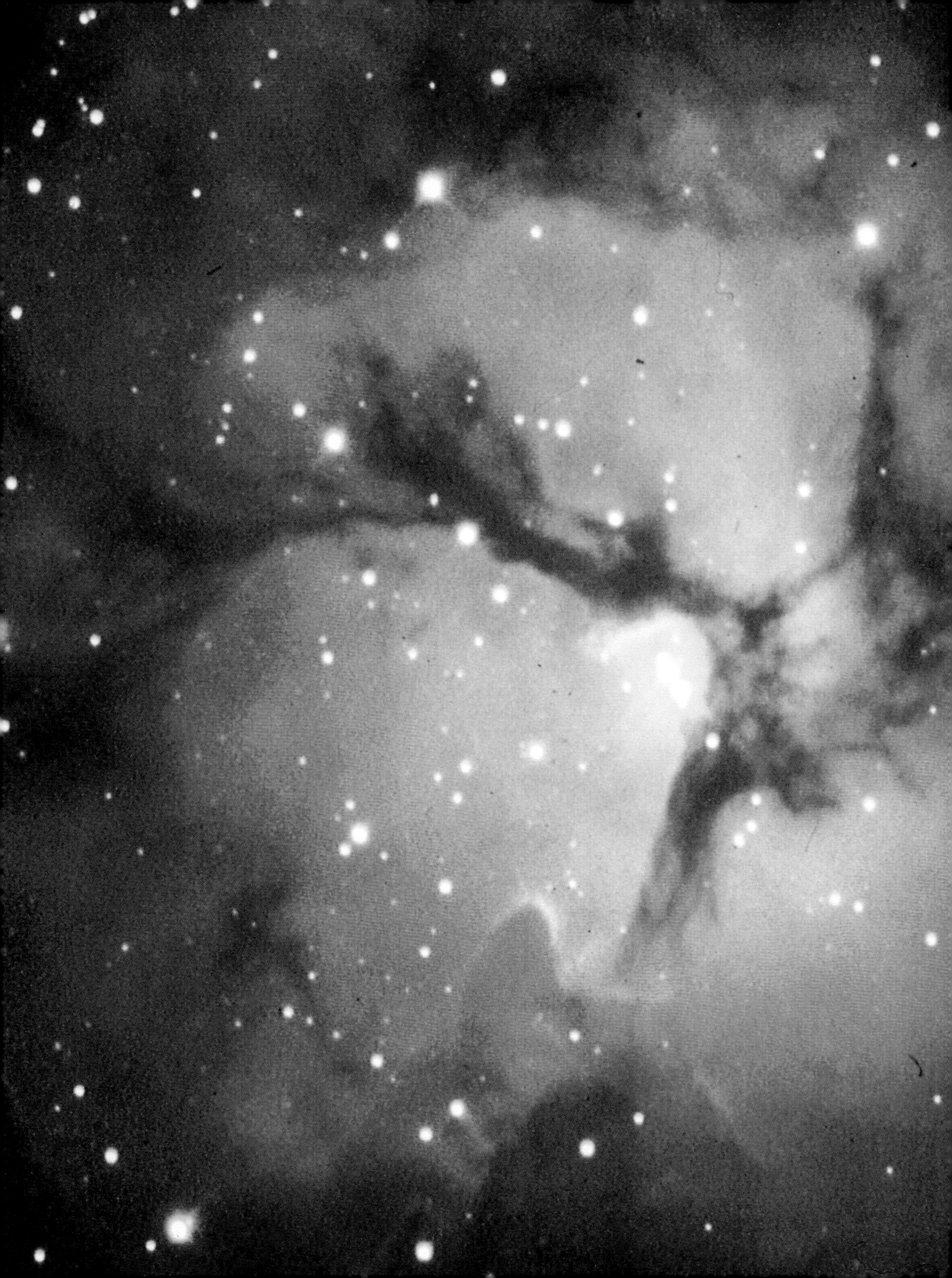

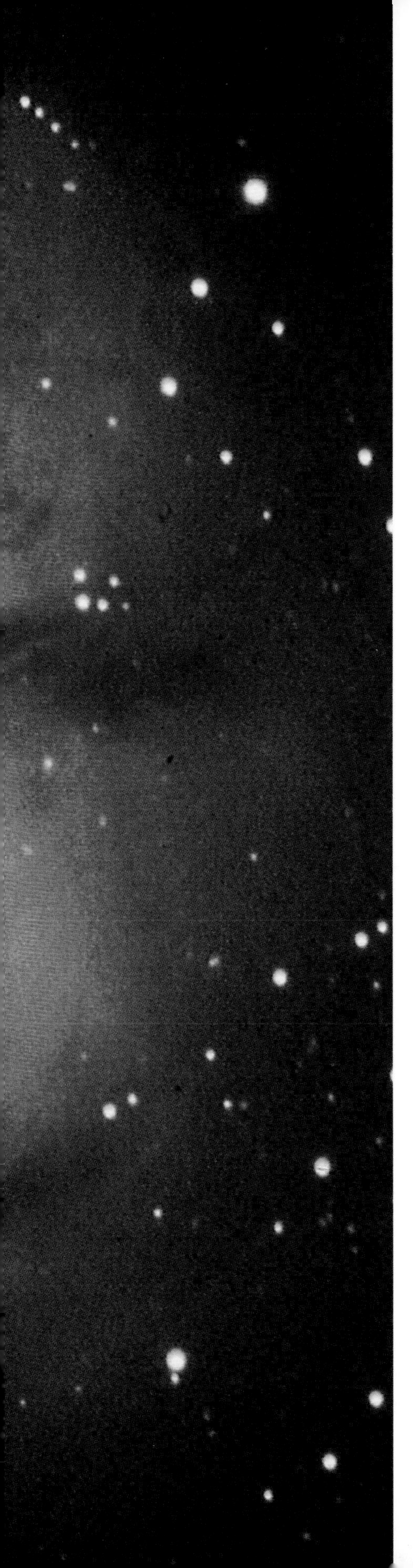

10 A Journey Through Time

We live a life of theory to connect a distant past with the future. To do so we search for the beginning of time and everything that has happened since that instant of creation.

We are a bipedal race on a small dense planet circling a minor sun that is part of an average galaxy. So we look back as far as sight will take us to capture in wondrous images the heritage of all things, and we measure past events because they always point to the future.

We are a family emerged from the accretion of glowing stardust into thinking, thriving, curious human beings, and we seek the thread to bind past, present, and the morrow.

We invented a time machine we call the Hubble Space Telescope to give us the clearest possible view of the universe.

Hubble peers with optical lenses and sensitive instruments through raging stellar dust storms, vast windy hurricanes of energetic particles, swirling bands of exploded stars, at blue giants and red dwarfs. It captures the light, energy, and motion of coruscating shock waves and X-ray beacons, and shows us the way into black holes where everything, including logic, vanishes. It does all these things and more.

Previous page: It took this image of the awe-inspiring, beautiful Trifid Nebula in Sagittarius 3,000 years to reach the telescopes of Earth, and to give us a view of what was happening in this birthplace of stars 1,000 years before the birth of Christ. By being 3,000 light-years away, the Trifid Nebula permits Hubble to see 3,000 years into the past. What Hubble is seeking is scenes 8 to 12 billion light-years away, images of an infant universe that existed long before our Earth and its solar system family were born. *(U.S. Naval Observatory)*

Right: Hubble sees the past by photographing "scenes of the universe" hundreds, thousands, millions, and most importantly, *billions* of light-years away.

By showing a graceful, spiral galaxy, as seen in this photograph, Hubble can help scientists plot the future of our own world.

Earth is located in a spiral arm of the Milky Way. *(U.S. Naval Observatory)*

But it is a time machine that cannot see the future. Its vision is locked into what is frozen in the past.

Hubble is an instrument of one-way time. It cannot even view the present, no more than any other machine or even the minds of men can. Only the past is available. Time rolls inexorably in a direction from *then* to *now*, but the very instant we think of an event, a person, a place, it has already become the past.

The past then serves us well.

Hubble permits us to separate the astrophysical catacombs by factual observation and measurement.

The journey of light and energy from across the universe to the huge telescope in earth orbit is one of strict limitation. Time is the master here. Nothing will reach Hubble's lenses without adhering to a universal law. Whatever has happened comes to Hubble, as it does to our eyes, with a speed limit of light at 186,000 miles per second.

For example, the tip of our nose or a hand an inch from our eyes are the only things we can almost see in the present. Even the image of a friend across the room is in the past. That image must travel across the room at the speed of light to reach our eyes.

All events, of any scale or distance, require time to reach observers. There is a parallel of an event and the time necessary to pass before that event may be *heard*. If you stand two miles from a great flash of lightning, you can see the flash almost instantly, in only .00000107 of a second. However, because sound in our atmosphere only travels at 1,000 feet per second, you will wait 10 seconds before the sharp crack and following rumble of thunder reaches you.

Light makes a round trip from Earth to the moon in 2.56 seconds.

Sound (if it could traverse vacuum, and does so only in theory) would make that same journey in 2,390,000 seconds—almost 40 hours.

Below: The images of the millions of stars we see in this photograph represent a journey of light across the universe to the Earth-orbiting Hubble Telescope. It is the very time needed for light to travel that explains the rule "The deeper you can see into space, the farther back in time you can go." *(U.S. Naval Observatory)*

Below: The RIng Nebula is located in the northern constellation Lyra, 5,000 light-years distant from Earth. Because it takes the scene you see here 5,000 years to travel across space to us, you are looking at an image older than the Pyramids. *(U.S. Naval Observatory)*

You are witness to a past event.

The words you are reading become the past as quickly as you read. They can be recaptured only by going back in time, even in seconds or milliseconds.

It is the *scale* of distance and time that is so different to our senses.

The diameter of our solar system can be measured two ways. It is nine billion miles from one side of the Oort Cloud to the other. Its diameter can also be measured as 13 hours—the time light requires to travel this distance.

When we capture light from stellar bodies, time and distance become one and the same. If a galaxy lies 45 million light-years from Earth, as does the giant NGC 4261 in the Virgo Cluster, we measure that distance in the time it requires light to journey from the bright galaxy to the lenses of Hubble.

If we measured in miles, we would likely run out of zeros the deeper we travel back through time.

Light travels a distance of 5.9 trillion miles a year. Obviously, multiplying 45 million years by 5.9 trillion miles becomes cumber-

some and, quickly, meaningless in the context of numbers we use in our daily lives.

So our yardstick is light-years in distance to other objects in the universe—and how they match events with which we are much more familiar.

We cannot, as yet, move back in time by any means other than light, whether it be for hours or years here on Earth, or billions of light-years through space. Time travel is the capturing of scenes and images by optical light or by instruments that measure energies invisible to our eyes.

Imagine your family gathered for a holiday, during which you will all be treated to home movies and videotapes collected as far back as movie film and videotapes were available. You watch the elders, grandparents, old friends, and children at play—including, most likely, yourself.

But none of the people in those films exists anymore, nor have they existed for a long time. Those who have died are either memories in your mind or memories captured on film or videotape. You watch yourself on film, scenes of your childhood, memories precious to you. Study that child carefully.

The child simply does not exist anymore. The child is a memory locked onto celluloid or magnetic tape. You cannot be both your adult self, *now*, and that child of decades before. One is memory only, the other is what you are today. And just as you have not for many years been that child that was, so tomorrow or next month or next year, you will no longer be who you are at this instant.

As it is with the memory of the child on film or videotape, astronomers peering through their telescopes are seeing objects in the universe as they were seconds, minutes, months, years, hundreds of years, thousands of years, millions of years, even billions of years ago.

For example, the Hubble team points the orbiting telescope at a ghostly smoke ring in space that cannot be seen with the naked eye. It is the Ring Nebula that began when the hot star puffed off its outer layer, creating a smoke ring 730 times the size of our solar system, and 5,000 light-years distant from our Earth.

Because it takes the light from the Ring Nebula 5,000 years to travel across space to Earth, Hubble is *not* taking a picture of the Ring Nebula as it is today. It is taking a picture of how it appeared 5,000 years ago—when the Great Pyramids were just being planned, the first stone blocks yet to be cut.

Right: Astronomers believe this Hubble photograph is a five- to seven-billion-year-old image of a universe that existed long before our solar system formed. Before Hubble is done, the huge telescope should take us back to the doorstep of creation itself. *(Space Telescope Science Institute and NASA)*

As Hubble continues its search through the universe, it focuses next on the Hercules Star Cluster 22,500 light-years distant. The huge telescope now sees images of stars that existed when the direct ancestors of man struggled to survive the fury of the last great ice age. *Homo sapiens sapiens* looked with awe at the night skies bathed in starglow. They had no idea that many of the stars they saw above them were images that had been traveling toward their eyes long before the Earth on which they stood was formed.

Because there is an image of almost everything that has happened in the past, moving at the speed of light through the universe, there is good reason to believe that one day astronomers will build a telescope that can see everything that has ever occurred since the first instant of creation.

Hubble likely will not be *that* telescope, but it will come fabulously close to the universe's beginning. The penetrating systems of Hubble have proven so spectacularly successful that astronomers now believe it will take them better than 90 percent of the way into the past—into and through an infant universe a billion years or less after the Big Bang.

Only 10 orbits after the repaired Hubble was put back to work in January 1994, its corrected and new cameras captured an image of our universe when it was 60 percent of its present age.

Astronomers studied the Hubble photograph they believe is a five-billion- to seven-billion-year-old image of a universe that existed long before our solar system was created—a picture showing several pairs of infant galaxies, a young universe boiling and expanding in enormous growth, rushing outward in all directions to develop into the starry wonder gracing our night skies today.

Before the story of human civilization is over, Hubble will take us back to a universe we cannot recognize. We will drift back to the doorsteps of creation, back to when the beautiful elliptical and spiral galaxies we see today were unwieldy clumps of protostar material, gathering into gravity-rammed spheres so massive they raged through stellar life, devouring their substance in violence that doomed them to premature explosions—stars being born and dying so rapidly they spattered the early universe like some fantasy of celestial fireworks. Out of chaos would come order.

The journey back starts here.

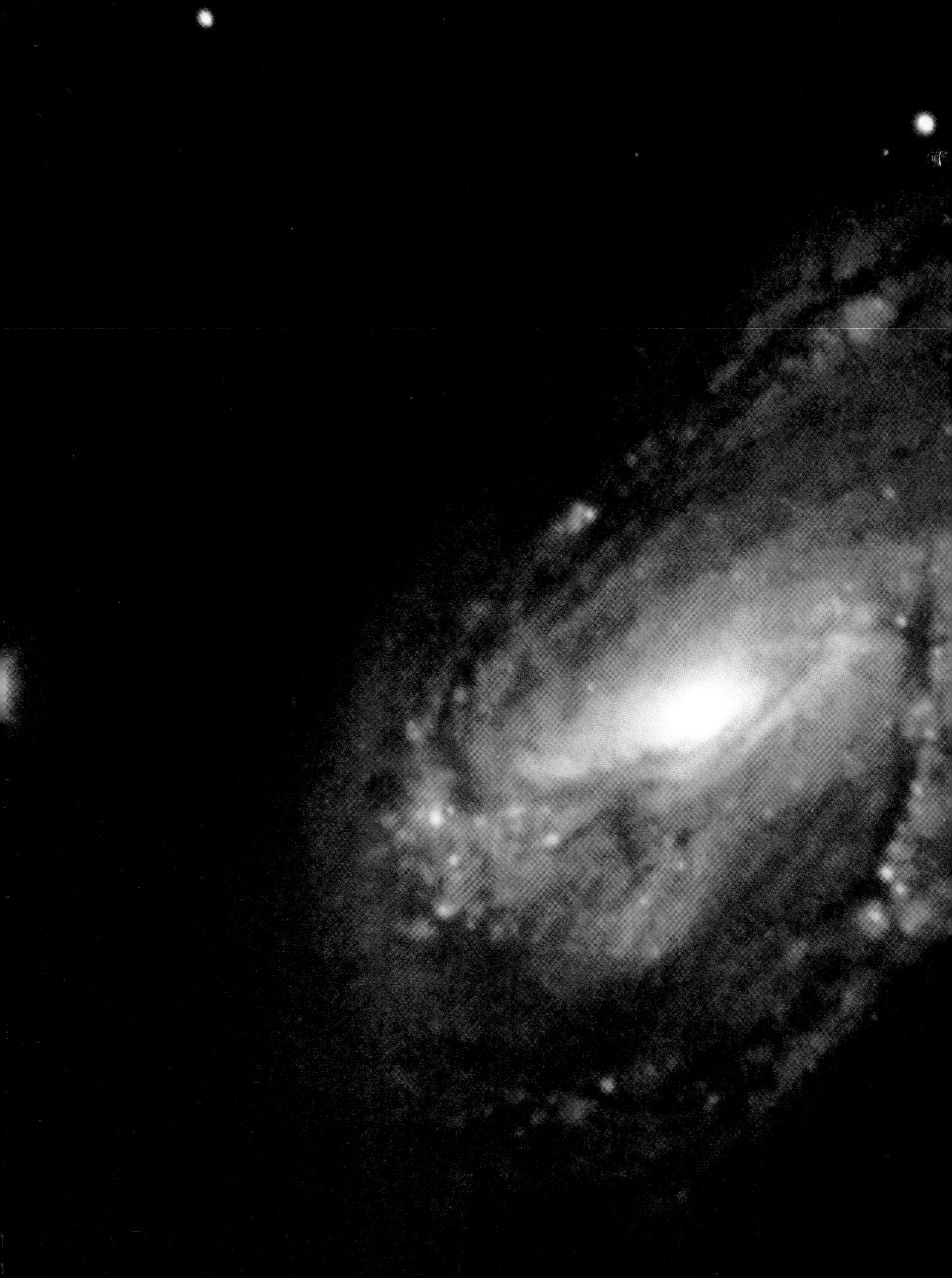

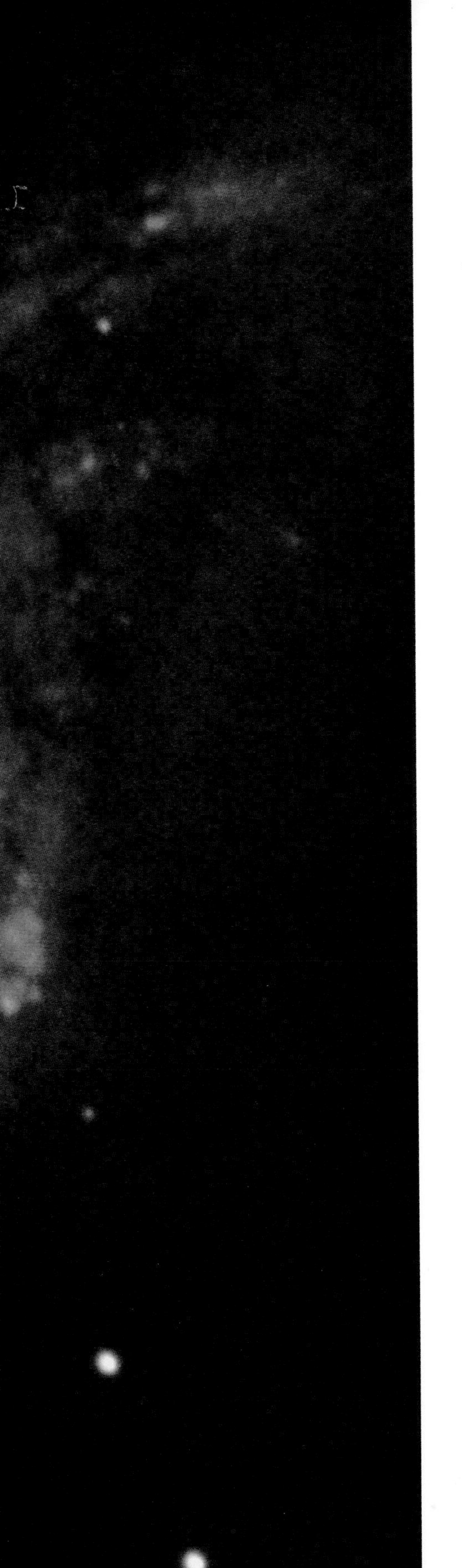

11

DOWN THE SPIRAL ARM

April 1990.

The Hubble Space Telescope is placed in orbit to begin piecing together the puzzle of the universe. Its mission is to tell us how the universe began, how it grew, how it is changing, and how those changes will affect our future.

We're told Hubble could focus on the universe's distant objects with ten times the seeing power of the best ground-based telescope. It would resolve blurry images across billions of light-years into pictures of great clarity. Closer to home, it would maintain surveillance of the planets of our solar system, adding steadily to knowledge of the outer worlds. Hubble would capture images of remote galaxies and nebulae not even visible to other astronomical instruments. We would learn vastly more about the size and shape of islands of galaxies, how and why stars form and die. We would *see* stars in their creation, stars exploding, galaxies colliding.

Above all else, stated enthused scientists, Hubble would push aside the shrouds of darkness that clouded the universe almost all the way back to the beginning of time. It would reach back for

Previous page: Our sun is at least 30,000 light-years distant from the galactic center of the Milky Way. *(U.S. Naval Observatory)*

Below: Galaxies have sometimes been called "island universes." In reality they are enormous spinning masses of billions of stars that also contain star clusters, nebulae, and vast amounts of cosmic matter. This NASA photograph of the giant Andromeda Galaxy (M-31) was taken by Mount Palomar's famous 200-inch telescope. *(NASA and Mount Palomar)*

8 billion, 10 billion, perhaps 14 billion years to show us a universe still in the throes of creation. As a time machine able to capture the past, so that we might learn more of how and why we came to exist, Hubble would be an instrument without parallel.

But there was something else, more meaningful than everything else put together.

The scientists who would tend Hubble—aim its cameras, turn its sensitive instruments this way and that, scour a universe seen only in ultraviolet light, then in infrared, in the spectrum of X rays and gamma rays, capturing the constant rain of cosmic rays, sensing all manner of subatomic particles—were of the same mind that we really could not anticipate what we *might* discover through the astonishing grasp of Hubble.

"As much as we know," said NASA's Chief Hubble Program Scientist, Edward J. Weiler, "we know far less than we would like to know. Hubble should find wonders far out in space of which we can't even conceive. That's the real payoff to everything we're doing."

He was right.

Our sun is in the suburbs of an island city of 400 billion stars. At least, that's the best count for the moment. The number of stars in our spiral galaxy were believed to number only several million to astronomers of old. As telescopes became more powerful, more stars swam into view. Millions became tens of millions and kept increasing to hundreds of millions; at last count scientists had settled on an estimated one billion suns. From one end of the Milky

Way to the opposite rim is a journey that requires light a hundred thousand years to cross. That's just the galaxy as seen in visible light. Once instruments seeing across the light spectrum became available, we found an enormous disk extending far beyond the stars we could see, a halo of faint stars immersed in dust and gas. Using the full spectrum, it takes light moving at 186,000 miles a second at least 400,000 years to travel from one edge of the Milky Way to the other.

The Milky Way is a great pinwheel of stars. In the center of the pinwheel an enormous bulge shines with such brilliance from stars packed closely together that it becomes a huge dazzling sphere obscuring all details. Here are found the great old stars of our galaxy in varying stages of their dying, adding to brightness by exploding by the thousands.

Through the spiral arms abound the younger stars of greater temperature and brilliance, among them our sun. Our solar system is not quite at the extremities of the pinwheel, but far enough out that the galactic center is at least 30,000 light-years distant. The entire mass—the island of stars of the Milky Way—like every other object in the universe, revolves about a common center, that dazzling bulge at the galactic core. Distances are so immense that the

Above: Twenty million light-years away in the constellation Canes Venatici, 60 billion stars fill the Whirlpool Galaxy, a smaller version of our own Milky Way. The companion galaxy, seen on the right of the Whirlpool in this photograph, is an extension of the spiral arm. Many spirals have similar, smaller satellite galaxies, and looking at the Whirlpool Galaxy in this photograph gives us an idea of what our Milky Way looks like from deep space. *(U.S. Naval Observatory and NASA)*

sun, speeding along at 130 miles every second, requires 250 million years to complete one revolution about the distant center mass of stars.

That nucleus of our galaxy was one of Hubble's prime targets for study. For several years, infrared and radio astronomy surveys had led scientists to suspect that at the very center, invisible to any kind of study from Earth, a huge star had exploded millions of years ago. In the blast of the supernova, the star began its last moments as a stellar furnace with a mass many times that of our sun.

A mighty shock wave streamed outward as the greater mass of the star collapsed upon itself. Swiftly, in a period of time yet to be reckoned, overpowering gravity rammed the star into a still-collapsing sphere of atomic nuclei jammed together. For an instant, the star trembled as a compacted sun only the size of Earth, but infinitely heavier. Still the collapse went on; now the sun had become a neutron star, barely a dozen miles in diameter, but with all its atomic structure reduced to neutrons packed tightly together.

The normal atom has a nucleus perhaps no more than one–one thousandths of the entire atom. The rest is empty space, the edges of the atom being the whirling paths of electrons. Electrons, protons, and neutrons jammed together as neutrons in the neutron star. A teaspoon of this neutron material would weigh billions of tons.

But this incredibly dense star, already spinning swiftly due to its contraction, remained in this form only for moments too brief to measure. So massive was the star when finally it tore itself apart, it collapsed in a microsecond or less.

About this stellar carnage, space was twisted and tortured; the very fabric of space-time, the substance of life itself, was shredded. The star was now so massive that the universe in which it existed could not sustain its terrible outpouring of force. Still the collapse went on until the entire star remaining from the explosion was reduced to a mathematical singularity—a point of space so small, so condensed, so intolerably dense that, in effect, it no longer could be seen.

It is invisible today, because it draws everything within reach of its gravitational grasp into a pit without a bottom. Star material, clouds, gas, dust, planets, anything and everything in space from individual atoms to huge stellar furnaces, pours from every direction into and down the black hole.

Not even light can escape. No gamma rays, or X rays, or light of any kind. At the heart of our Milky Way a hideous cannibal has savaged the space and time about itself and now is bent upon the

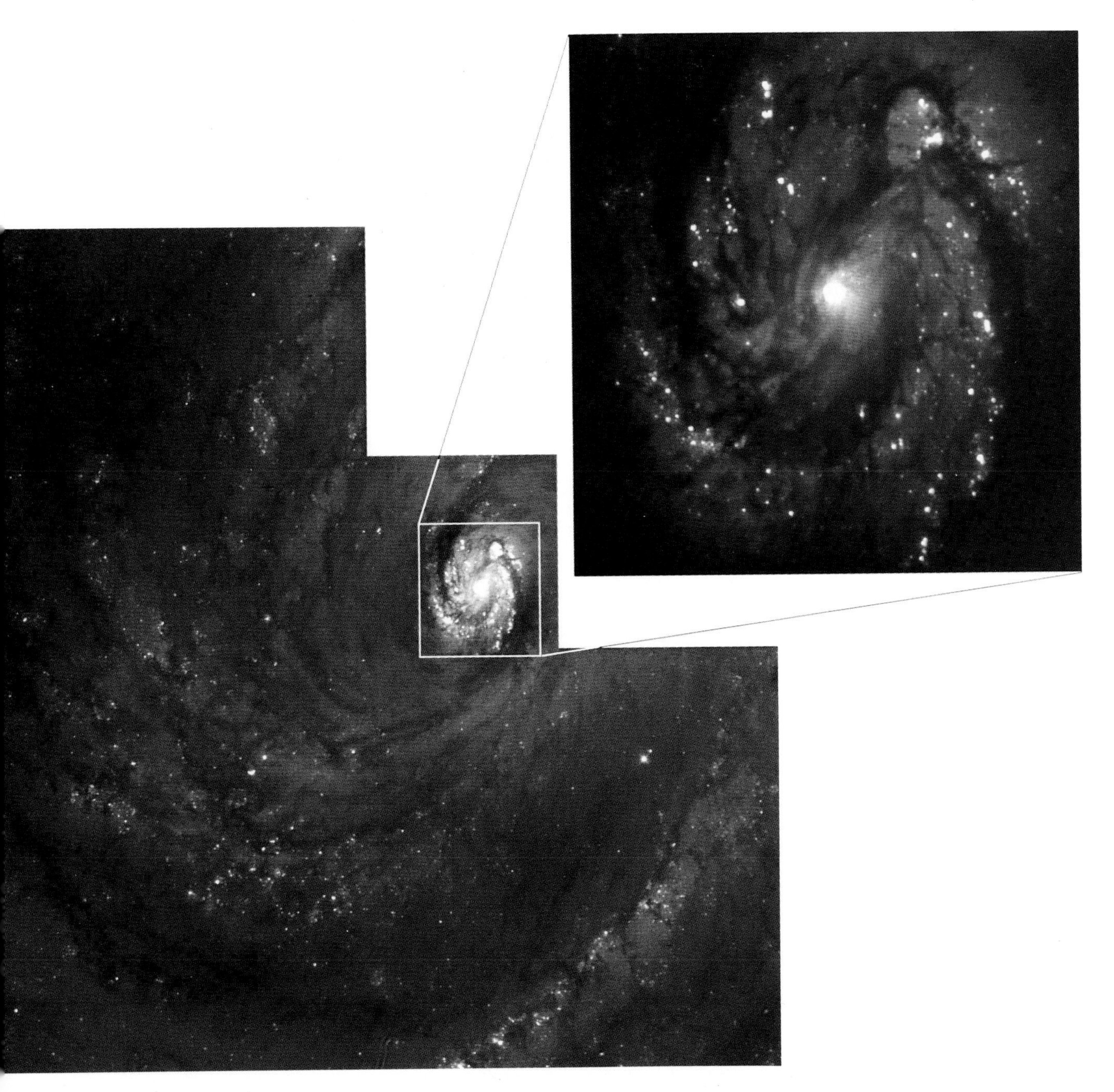

Above: A Hubble photograph of the M-100 Spiral Galaxy in the Virgo Cluster. Scientists had suspected for several years through infrared and radio astronomy surveys that at the very center, invisible to any kind of study from Earth, a huge star exploded in the core of the Milky Way millions of years ago. *(Space Telescope Science Institute and NASA)*

ingestion and destruction of everything it can affect. The only way Hubble knows this monster exists is its effect upon surrounding space. Material drawn into the black hole spirals inward. It is whirled about the ultimate engine, faster and faster. Now it forms an accretion disk, like an enormous platter surrounding the darkness that can never be seen. Faster and faster, until the material sweeping inward moves so swiftly it glows from its motion. Some of its energy is whipped away before it can be sucked into the black hole.

This is like a line of ice skaters circling one person at the center of a long line. The skater in the middle turns very slowly. The

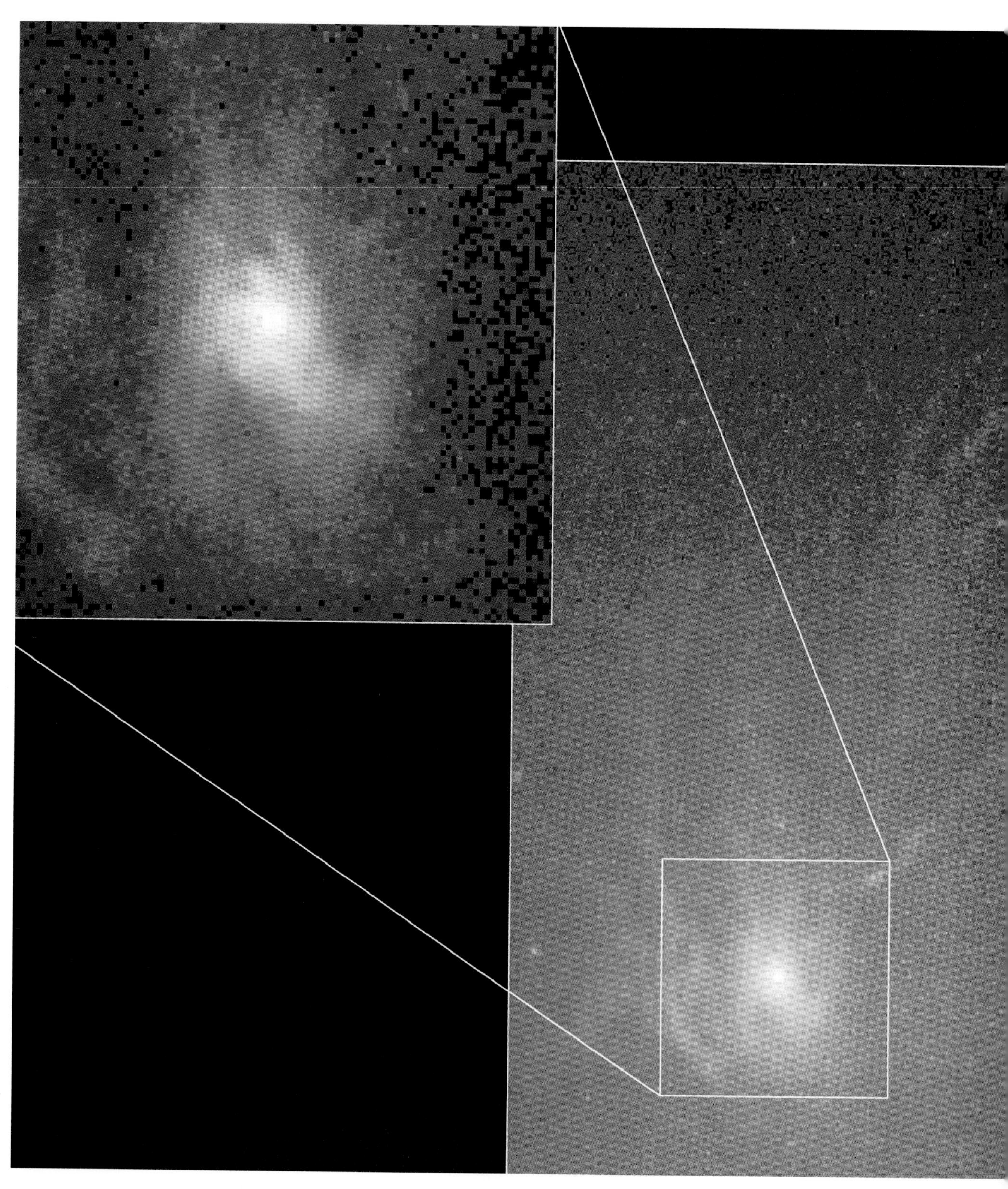

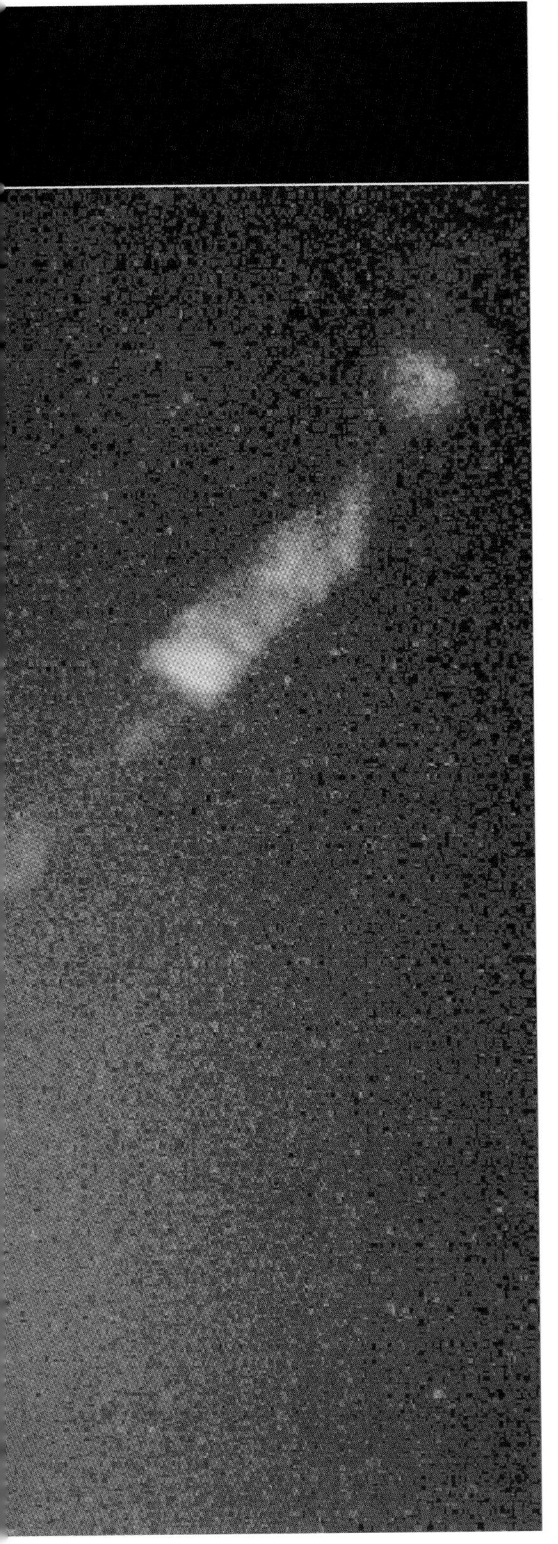

Left: In this photograph, Hubble measurements show the gas disk in the Nucleus of Active Galaxy M87 50 million light-years from Earth is rotating so rapidly that not even light can escape from this massive black hole at its hub. Many astronomers feel there is such a black hole at the heart of our Milky Way, a hideous cannibal. *(Space Telescope Science Institute and NASA)*

farther out from the center, the faster each skater must go. The last skater races breathlessly simply to keep the line straight—and, finally, can no longer hold on to the arm of the next person inward on the line.

This last skater goes whirling off, away from the line. This is the energy from the edge of the accretion disk, X rays, waves of gravity, glowing light in the ultraviolet and infrared, gamma rays, hurled outward in a frenzy of energy—which is what Hubble saw and scientists interpreted, which told them of the ultimate violence at the heart of the Milky Way.

At this speed, any mass, be it a pebble or a mountain, reaches a mass that is infinite. It cannot move any faster than it is going, no matter how much energy it expends. This is the Holy Grail of all science.

Except the black hole. According to every physical law we know, a black hole cannot logically exist. Its gravitational attraction is so great that inward acceleration of matter, which includes the photons that make up light and are quanta (matter), exceed the speed of light. The cannibal at the heart of our galaxy, it seems, pays no heed to the laws of reality. *Everything* vanishes down its maw.

The next question is, obviously, where does all that stuff go? Where does stellar material, pure energy itself, emerge? *If* it emerges, that is. Now we're into the area of guesswork, of endless and argumentative theories, because speculation is all science has when confronting this aberration of reality.

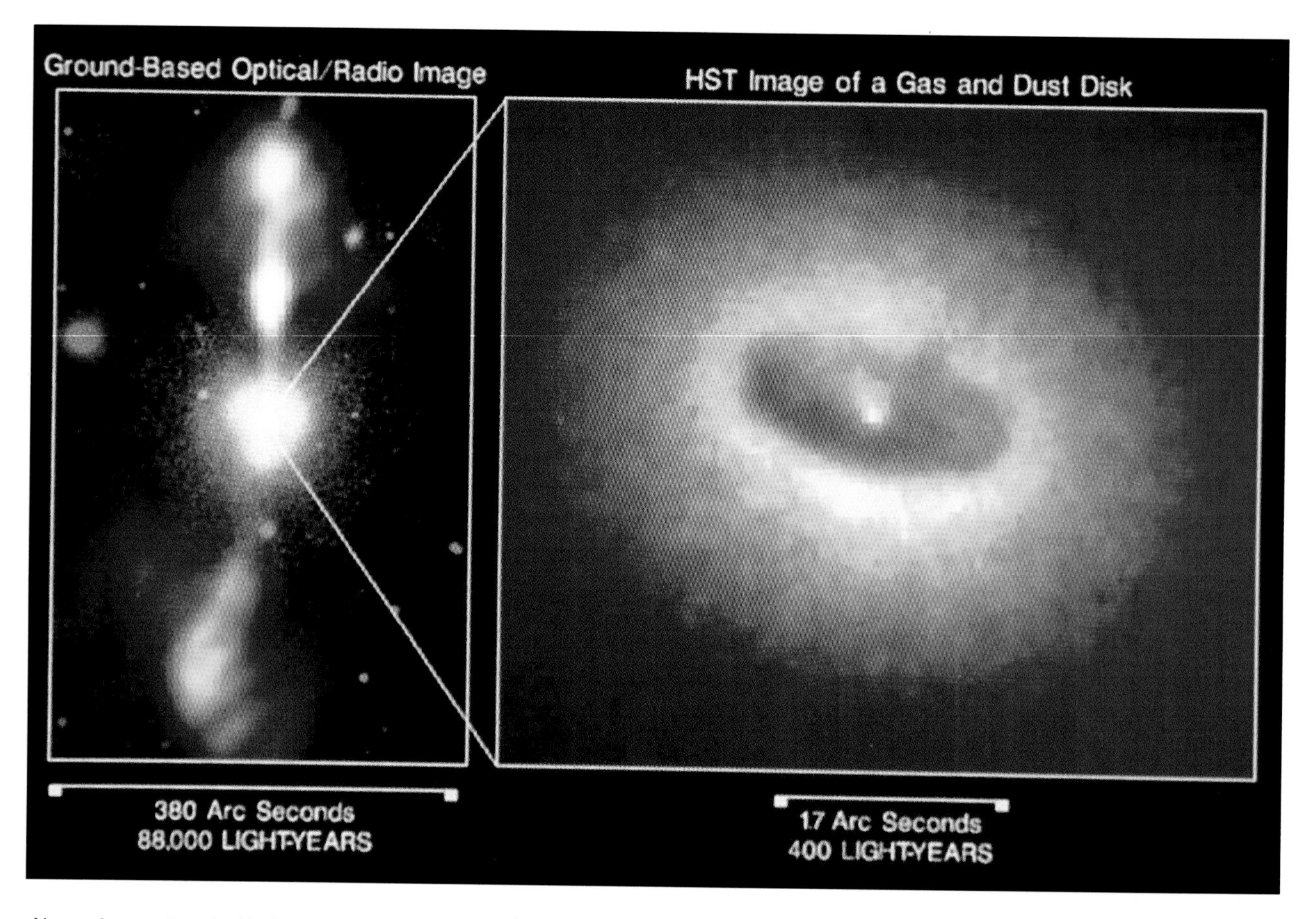

Above: As seen here in this Hubble photograph of the core of Galaxy NGC 4261, material drawn into a black hole spirals inward. *(Space Telescope Science Institute and NASA)*

But the sudden violence that produces a black hole, when just this one star overwhelms the brilliance and energy of a billion suns, while invisible, remains where it was created. It is now pure cannibal.

The accretion disk surrounding the invisible core now whirls faster than a million miles per hour. Still the heat of indrawn material increases, sending out dazzling jets, fountains, and clouds of energy that we detect with our instruments. They tell us that whatever is going on at the galactic center continues to get worse.

Cannibal is no metaphor! It is the descriptive phrase that scientists have adopted to describe the voracious appetite of the monster ingestion at the heart of the Milky Way. Referring to the size and estimated mass of our galaxy, scientists judge that the black hole of the Milky Way has a mass equal to 10,000 suns. It is a gravity engine that exerts so much force for millions of miles in all directions it is now accepted as the power source of our active and energetic island of stars.

Of a black hole specifically, Astronomer Bruce Margon of the University of Washington states, "It will keep eating as long as a meal is around." The gas and material within reach of the gravitational pull "is lunch. It's going to fall into the black hole and be gone."

Scientist Richard Harms of the Applied Research Corporation sounds a less sinister note. "It [the black hole] has an infinite appetite but it can't hunt," he explains.

Perhaps so, but it doesn't have to "hunt." Nothing in space is ever static; nothing is ever still in relation to other objects and material. Everything moves, everything rolls in circles or is hurled outward from stellar violence of many kinds. Everything continues to fall within the massive attraction of the black hole. Sooner or later, virtually everything will fall within the clinging sweep of its gravitational field and be drawn into an abyss impossible to measure, or depict, or perhaps even to understand.

Hubble has given us another look into the ultimate abyss. Calculations of the effect and reach of the Milky Way's black hole lead to the conclusion that the monster has already consumed, utterly and without a shred of anything left over, several *billion* stars.

Before the age of Hubble, we had long regarded our galaxy as a spinster aunt of stellar decorum. Our island universe wheeled about in stately motion, stars arrayed neatly in long curving lines, their actions and energies quite predictable and well-behaved.

Hubble proved us so wrong!

12 THE WONDER OF IT ALL

In many ways the Milky Way is characteristic of millions of other galaxies. They are all islands of stars and accompanying gas, dust, molecules, nebulae, stellar detritus, planets, and smaller rocky bodies. They all revolve about a central core dictated by density-spawned gravity. Many are smaller than the Milky Way; many are larger. Some have trillions of stars, others much less than 100 billion; our Milky Way has 400 billion.

In all these island universes, stars almost continuously are born, blaze fast or slow, and die. Some stars die quietly, withering away, depleted of life until they no longer cast off energy—the flotsam of the universe, dark and forgotten bodies that once were warm and bright. Others stars explode with galaxy-shattering violence as supernovas. Sometimes the supernova fails in the midst of its own final detonation; the building blast becomes an astonishing flurry of stellar stuttering, ragged and out of control, releasing mass and energy in dazzling bursts of X rays and gamma rays—unfortunately, lethal to whatever inhabited planets are within reach of such deadly cosmic rain.

Through our galaxy and, as far as we know from what all our

Previous page and above: As seen here in these two photographs by the U.S. Naval Observatory, the galaxies M-64 and NGC 4565 revolve about a central core. *(U.S. Naval Observatory)*

telescopes and instruments tell us, most if not every galaxy is burdened with the rogues of creation—the cannibalistic black holes of so much density and power that the very fabric of space-time is shredded and billions of stars consumed—to vanish forever.

The Hubble science team has been on the hunt for another enigmatic source of destruction—black holes so small they are as tiny as a pinhead, yet exert more energy than full-sized stars. Theory suggests these wreak havoc through star systems, plunging completely through a planet or star, leaving wreckage as they hurtle onward. It is an area of mystery; the micro-sized black holes appear to be real, and finding them may well be the most difficult of all assignments for Hubble.

But there is no doubt of the extraordinary success Hubble produced when the telescope's Planetary Camera was trained on Nebula NGC 2440 within the Milky Way. Astrophysicist Dr. Sally Heap of NASA's Goddard Space Flight Center and her team of astronomers were given the task of capturing an image of one of the hottest stars of the universe.

The problem that even the most powerful ground-based telescopes could not overcome was twofold. Pictures made by these observatories were doubly smeared, first from the thick and turbulent atmosphere of Earth, but especially from the dazzling glow of the nebula. Hubble was aimed straight for the central star of the sight-shrouding glowing cloud. Through use of the Planetary Camera and computer-enhanced resolution, Dr. Sally Heap's team managed to separate the light from the star at the core of the nebula. The Hubble team scored an incredible bull's-eye with their program. Photographs showed clearly a single bright white source in the center of the images. There, a star of fantastic energy blazed with a surface temperature of more than 360,000°F—compared to the surface temperature of our sun at only 10,000°F. Until the Hubble images of the NGC 2440 nucleus, many scientists doubted

Below: This Hubble view of the barred spiral Seyfert galaxy NGC 5728 reveals a spectacular bi-conical beam of light that is ionizing the gas in the galaxy's core that might contain a supermassive black hole. *(Space Telescope Science Institute and NASA)*

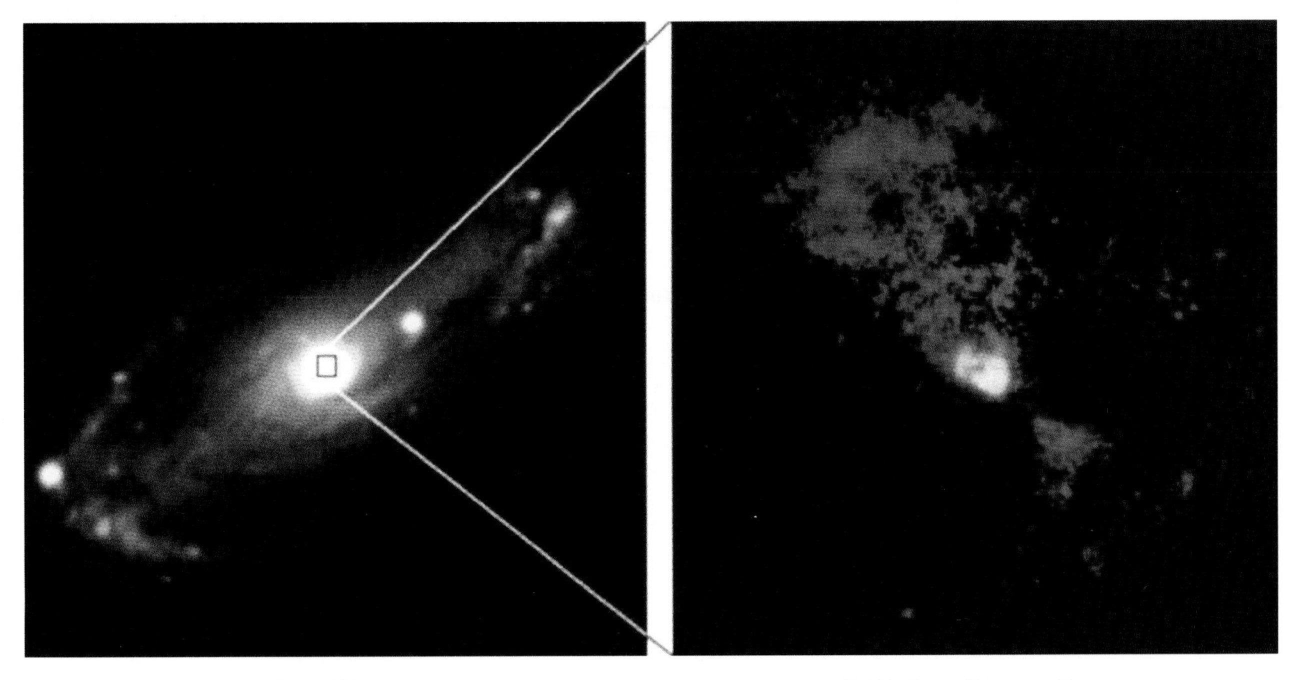

Ground View

Hubble Space Telescope View

that any star could burn at this extraordinary temperature without self-destructing. With that single image Hubble wrote a new page in the astronomy texts.

In a region of 65,000 cubic light-years about our sun are to be found at least 12 stars of the same general size, temperature, and energy levels as the sun. These are future targets for Hubble, for they are all candidates for solar systems. If they have planets that fall within what is considered a safe enough distance from the star to support life, as does Earth from our sun, they may well be the first extraterrestrial worlds on which we will detect intelligent life.

Within 13 light-years of our sun are 23 stars that astronomers believe fit within the profile of suns likely to have planets in orbit. The stars and their distances from Earth are:

Proxima Centauri	4.2 light-years
Alpha Centauri	4.3 light-years
Luyten 726-8	5.8 light-years
Barnard's Star	6.0 light-years
Wolf 359	8.1 light-years
Lalande 21185	8.3 light-years
Ross 154	9.3 light-years
Procyon	10.4 light-years
Ross 248	10.4 light-years
Eridani	10.7 light-years
61 Cygni	10.9 light-years
Ceti	10.9 light-years
Ross 128	11.2 light-years
Indi	11.3 light-years
Struve 2398	11.4 light-years
Groombridge 34	11.6 light-years
Lacaille 9352	11.9 light-years
Cordoba Vh. 243	12.4 light-years
B.D.-5 1668	12.4 light-years
Ross 614	12.5 light-years
Cordoba 29191	12.6 light-years
Kruger 60	12.7 light-years
van Maanen's star	12.8 light-years

If we could journey inward from Earth toward the galactic center, as do the optics and instruments of Hubble, we would never again look with disinterest at the skies filled with stars. Familiarity breeds, if not contempt, a complacency that all is the same, unchanging, and we see stars across a void that stretches

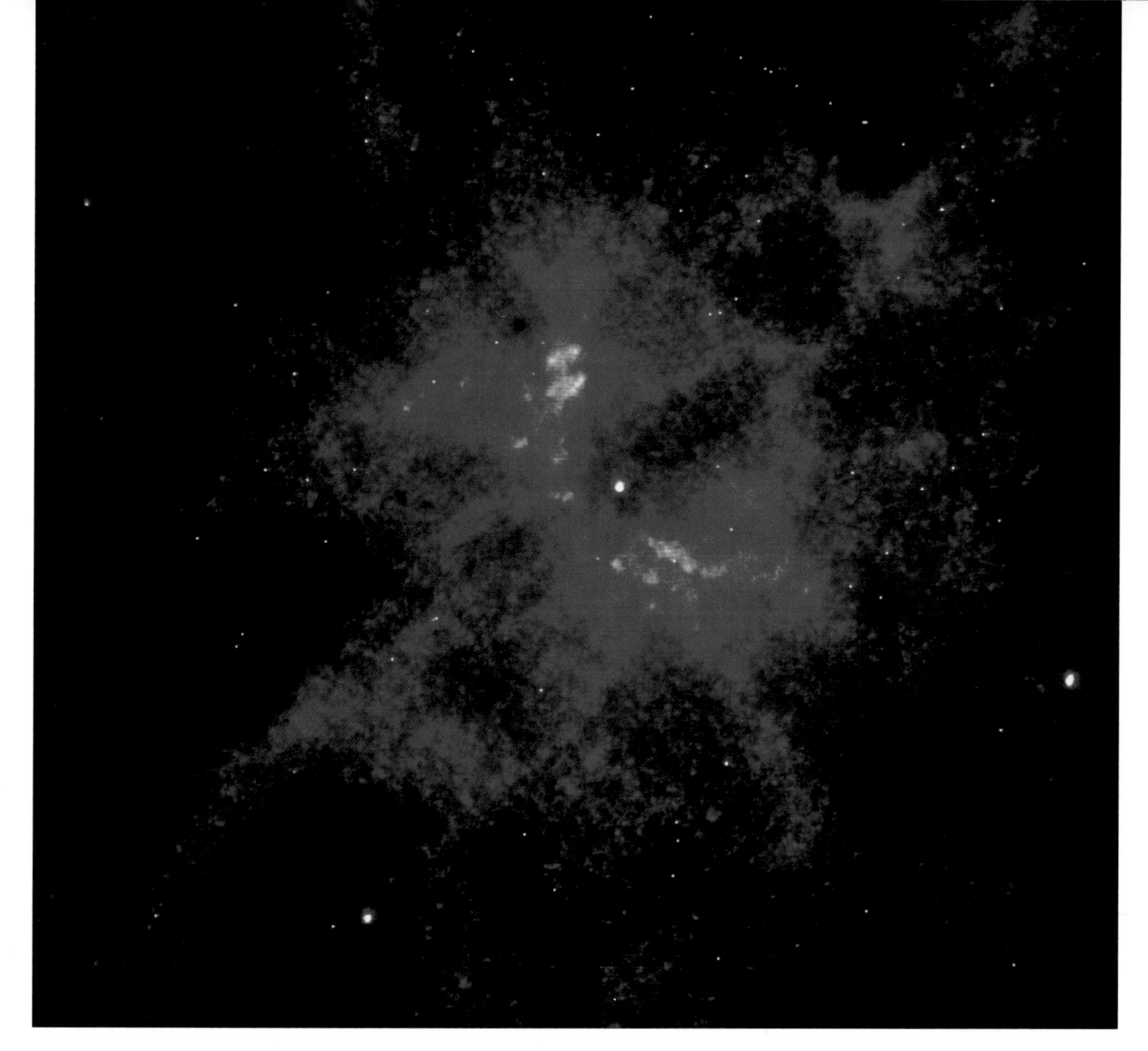

Above: A Hubble photograph of a single bright white source in the center of one of the hottest known stars, the central star of nebula NGC 2440 in our Milky Way. *(Space Telescope Science Institute and NASA)*

forever. One need only study photographs of star fields to see that every possible face of the image is swarming with stars, so many they are uncounted, so many they blur into a single picture-wide source of light.

In those pictures space *is* crowded with stars, so many suns jostling one against the other there hardly seems room for anything in between. The problem with such viewing is that we are looking at a flat, two-dimensional image. There is no sense of the *depth* of space, or that space and all it contains radiates outward from where we are, in every direction, a bubblelike diaphragm that goes on, as best we know, forever.

Imagine, then, that we depart Earth and rush outward to the outer rim of the Milky Way. Here the stars are sparsest, flung away from one another. Yet even with this separation there are so many suns they seem like luminescent froth riding breaking surf on a beach.

Above: As seen in this U.S. Naval Observatory photograph of the central Milky Way, Earth is a small dense planet, circling a minor sun that is part of an average galaxy. *(U.S. Naval Observatory)*

Now inward we go. Something unexpected happens—just as long as the journey is made with great speed to cover the enormous distances. Stars flash by, some above and others below, but always on both sides and ever denser far ahead.

Now the amazing complexity and variety of the Milky Way appear in dazzling colors and shapes. Nebulae glow as slowly expanding clouds of fireflies. Long jets of star matter arc outward, some straight as arrows, others curving and spiraling, having left far behind the star that spat them outward. Every color and shade and hue imaginable fills the view in every direction, but any traveler making so wondrous a plunge must be reminded that it is only the speed, much faster than light itself, that crowds the spiral arm with numbers far more than could be counted. If this plunge inward were made at only the speed of light, nearly six trillion miles in one year, the voyage would take 50,000 years just to reach the core of the Milky Way.

But imagination is the vessel within which we make this flight. We travel far faster than the speed of light so we might encompass even a smattering of the wonders of this galaxy. If we continue straight ahead, dodging only those stars or planets that might interfere with movement, we pass through the outer atmospheres of red giant stars so huge they could swallow most of our own solar system. They are real stars, burning at their cores, but expanded like the flimsiest of outer onion skins, the stellar atmosphere less than one–two thousandth the pressure of air on Earth.

Another surprise carries all through the inward rush. For every sun sailing by itself through space, there are two or three pairs of stars, binaries revolving about a common center. Sometimes there are three, a dance of suns in a slow-motion stellar waltz.

In every direction lie the clusters, golden chandeliers of a billion pinpoints of light, thousands of light-years across. The sparkling dots of stars at the outer edges mark the beginning of density. Plunging through the stellar thicket always means journey's end in a fantasy of stars jostling and rushing about one an-

Below: A 10,000-light-year-long jet of plasma photographed by Hubble. *(Space Telescope Science Institute and NASA)*

other, packed so tightly there is no way to avoid collisions. Stars meld one into the other, casting off violent streams of energy, flashes of visible light and unseen blasts of X rays and subatomic radiation that strike gaseous clouds to set them ablaze with sharp colors and soft blending of hues.

It is in constant activity, this galaxy of ours. Stars of every color, even stars twisted and bent by forces we do not understand, so that instead of spheres they have become the shape of fat and swollen footballs.

For a moment we pause to consider a yellow star burning steadily, a twin to our own sun. Not until Hubble penetrated distance and vast clouds of dark matter did we learn that the universe is a place of reincarnation. The word *reincarnation* itself seems utterly out of place and time in the stellar arena, but it is true. By studying the activity of the sun—how it burns, consumes its energy, examining the elements cast off, measuring everything—the astonishing reality emerged that what is our sun today is a star three times removed from its original form.

At first it was a star that grew to mighty size, bursting with energy, so massive that finally it exploded. Shells of gas rippled away in all directions, clouds rushed from the terrible blast. Then a following shock wave caught the gaseous halos, smashing the gaseous material into glowing froth. The halos collapsed, then began to assemble from the force of gravity demanding accretion until, finally, the gaseous material all gathered into a new mass. Again the stellar birth process began: Mass compacted, density grew, temperatures soared, and hydrogen fused into helium—and the star was born anew.

Below: Hubble took this photograph of the M-87 jet leaving its star. *(Space Telescope Science Institute and NASA)*

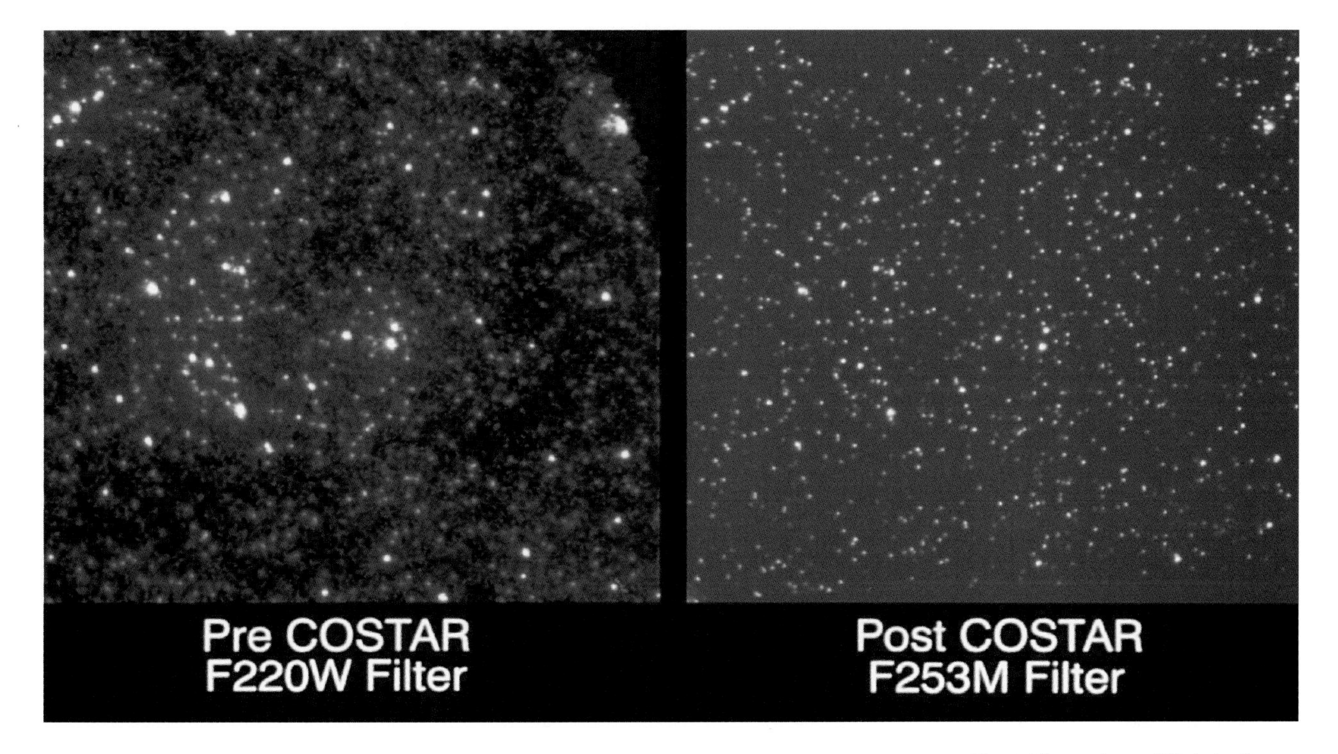

Above: The nucleus of Globular Star Cluster 47 Tucanae. *(Space Telescope Science Institute and NASA)*

The cycle went once more through billions of years. Growth, blazing heat, runaway fusion, and another explosion. Our sun is the third time around, a process that goes on all the time in this wonderland of galactic stuff.

By now it must be obvious that within a galaxy such as ours all space is packed with stars, shoving and bursting and wheeling through invisible lines of their orbits.

And that, as Hubble has confirmed from earlier theoretical observation, is a facade of our universe.

When we stand beneath a clear sky at night, the air cold and as still as glass, the stars appear to burst forth in numbers so vast they cannot possibly be counted.

Reality is otherwise. If you had the time and the patience, you could count every star visible to your eyes, but you would never count more than 3,000 stars.That is all our eyes can encompass.

There is an overwhelming, even awesome, impression that we see vast distances. We know the universe extends outward for billions of light-years, but you never see a single star, a single point of light in the heavens, that is not part of the Milky Way. Constellations, clusters, nebulae, everything that our eyes can discern is within. Only through the reach of optical instruments, especially those of Hubble, do we manage to break the bonds of distance, and capture the light that has been rushing toward us since the beginning of time.

The incredible numbers, the billions of billions of stars contained only within the Milky Way, are a constant paradox. A glance at the night heavens reveals the immense crowding of stars. A

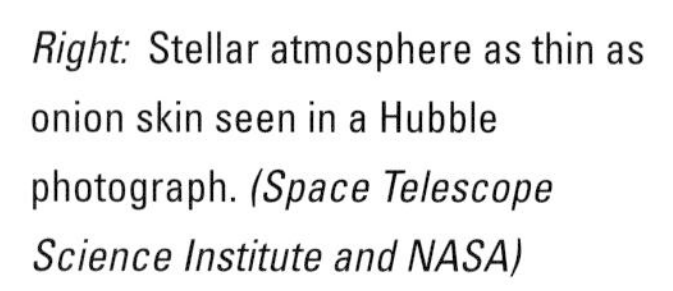

Right: Stellar atmosphere as thin as onion skin seen in a Hubble photograph. *(Space Telescope Science Institute and NASA)*

study of photographic images pushes away black and leaves only a total smear of starlight.

But space is so vast that not even trillions of stars can produce more than local gatherings. Everything else is devoid of the stellar furnaces. Space essentially is mostly empty.

Years ago, one of the all-time great astronomers, Dr. Fred Hoyle, established a numerical estimate of what he considered to be inhabitable planets in only the Milky Way. At the time he presented his numbers, astronomy judged the number of stars in our galaxy at only one-fourth of what we know today. To Fred Hoyle, the Milky Way contained 100 billion stars.

Out of that vast number, he judged, only a fraction of those stars would have planets. That "small fraction" in his calculations yielded at least one million planets.

He added that if only one in every ten planets discovered could support life—human life, he stressed—this meant that in the Milky Way alone, there must be a hundred thousand inhabited worlds.

Yet astronomical calculations show that if the Empire State Building in New York City, 102 stories tall, were emptied of everything save two specks of dust—it would be more crowded than is space with all its stars.

The search for planets about other stars has been ongoing for decades. But not until modern astronomy—optics and spectroscopic instruments, devices able to scan the universe in all wavelengths of the electromagnetic spectrum—have we achieved positive results. Astronomer Alexander Wolszczan at Penn State

University confirmed the existence of several planets in the Virgo constellation of the Milky Way. Fifteen hundred light-years (seven quadrillion miles) demands incredible patience in trying to detect a planet-sized object, usually masked by the brilliance of its host star. By measuring gravitational radio signals from a collapsed star, Wolszczan found three planets, two about 24,000 miles in diameter, the third only 2,200 miles in diameter.

Enter the Hubble Space Telescope with direct observational results. In late 1993, reports that Hubble research teams were on the brink of extra-system planet discovery sent excitement through the entire field of astronomy. New evidence confirmed that stars by the millions were breeding planets. The Hubble teams, their sights trained on the overwhelming beauty of the Great Orion Nebula, presented images of 15 new stars to the delighted Hubble scientists. Fifteen hundred light-years distant from Earth, behind the sword of Orion, shone the new stars.

The dazzling suns, some dimmed by the gaseous filament of the nebula, weren't the star attraction of the moment. But about each star there extended a flattened disk, unquestionably the same kind of interstellar gas and dust from which all planets must begin their process of accretion, eventually to become the very types of planets in our own solar system.

"Protoplanetary disks" was the name given to the formations about the Orion stars. "These images provide the best evidence for planetary systems," stated the scientist who made the discovery, Astronomer Robert O'Dell of Rice University.

The find by O'Dell's search team was entirely a different matter from Wolszczan's discoveries. Here was *visual image* confirmation of forming planets. And, to the later delight of the O'Dell team, they pointed out that their pictures were taken with the Hubble Space Telescope prior to its repair.

The former prediction by Fred Hoyle, British great of astronomy, assumed startling importance. Working on their own, researchers reported their studies show, conservatively, that some 10 percent of stars must be accepted as having planetary companions. The ratio of one in every ten stars hosting planets proved right on the mark set by Hoyle decades earlier. Further research produced new predictions—mainly that if Orion was an accurate benchmark, the odds of finding galactic planets were getting better all the time.

In addition to the Wolszczan studies, and prior to the Hubble photos, ground-based telescopes had found clear evidence of protoplanetary disks spinning about at least four planets through the Milky Way. Four out of 400 billion isn't even a dent in the possibil-

Below: Our sun today is a star three times removed from its original form. *(Space Telescope Science Institute and NASA)*

ity factor of planet-finding, but then, the ability of telescopes to resolve distant images left much to be desired. And those disks left most astronomers dismissing the protoplanet possibility as extremely slim. The disks were old material swept up by gravitational vacuum cleaners, rather than forming when the stars themselves accreted, as in the case of our own solar system.

Below: In the constellation of Ursa Major are very interesting twin galaxies. As seen in this photo, one is bright orange, the other galaxy is blue. Notice the fine wisps at the edge of each galaxy. There is evidence the bright orange star we see here is exploding. *(U.S. Naval Observatory)*

There lay a critical point in the Hubble photographs, which is what ignited astronomers' hopes that we'd finally latched on to real planets in their making. Gas and dust that collects about a young star spins with great speed—much too fast to be sucked down into the gravity traps of the stellar parent. This is proof to astronomers that both star and disks formed at the same time, and the disk of material as it spun, safe from the cannibal appetite of the host star, would form eventually into clumps of matter, which would then begin to draw in additional material from their own growing gravitational attraction. Add a few billion years to the process, and planets and moons would emerge out of the disk.

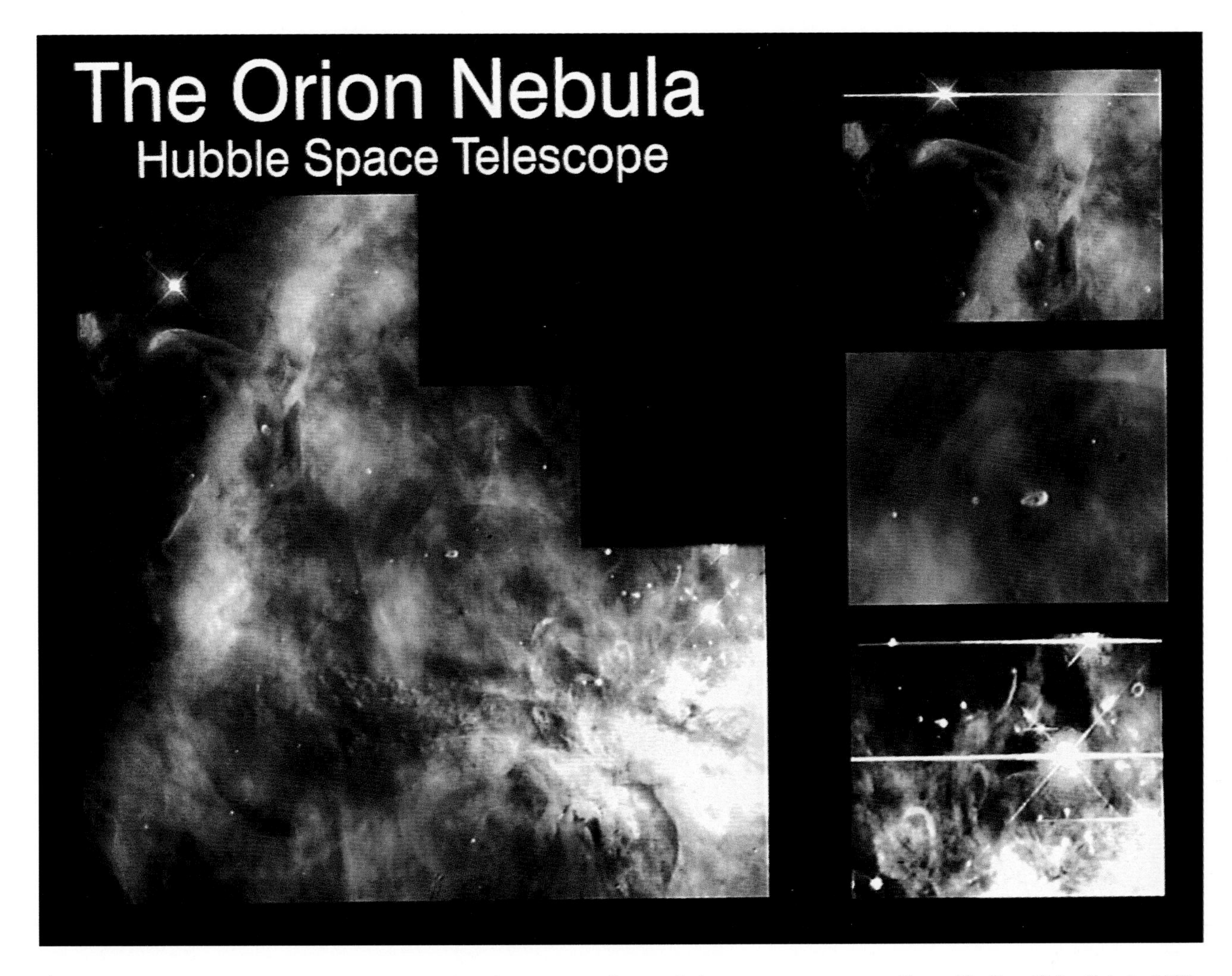

Above: The Great Orion Nebula, 1,500 light-years distant from Earth. *(Space Telescope Science Institute and NASA)*

After these epochal pictures, scientists went through impatient nail-biting while NASA prepared for the repair mission to correct Hubble's cameras. The fateful day came and went, and O'Dell's team trained Hubble on the same region of the Great Nebula in Orion to find the protoplanetary disks captured during earlier observations.

The results were astounding. The photographs confirmed that at least half of all the stars in the area of the nebula were showing signs of forming planets—and, in a sweeping vindication that going back in time establishes contemporary conditions, that we were watching the equivalent of how planets formed in our own solar system.

"This provides strong proof that protoplanetary disks are a common part in star formation," said Robert O'Dell. O'Dell's team studied at least 100 stars out of a star-birthing region of more than 3,000 suns. Of the 100 that were captured in images by Hubble, at least 56 of these were what O'Dell christened "proplyds"—very young stars that hadn't yet blazed through their first million years of life.

Below: Hubble sent back this picture of a star-forming region of space called the 30 Doradus Nebula. *(Space Telescope Science Institute and NASA)*

O'Dell again referred to the similarities of accretion growth between the Orion region and our own solar system during its earliest age. "The most exciting result is that at least half of these cool young stars are each surrounded by their own small clouds with dimensions about five times that of our [present] solar system."

It turns out that Fred Hoyle was indeed a prophet with both honor and astonishing accuracy in his predictions.

Searching for planets both within and beyond our galaxy has already produced exciting—but hardly unexpected—results. Nothing more than the mathematical odds greatly favor planetary development throughout the universe. In the hundreds of billions of galaxies and the multiple trillions of stars, to accept that there is only our one aging yellow sun with nine planets as the sole source of life—let alone intelligent life—is hubris. It is not a case of the

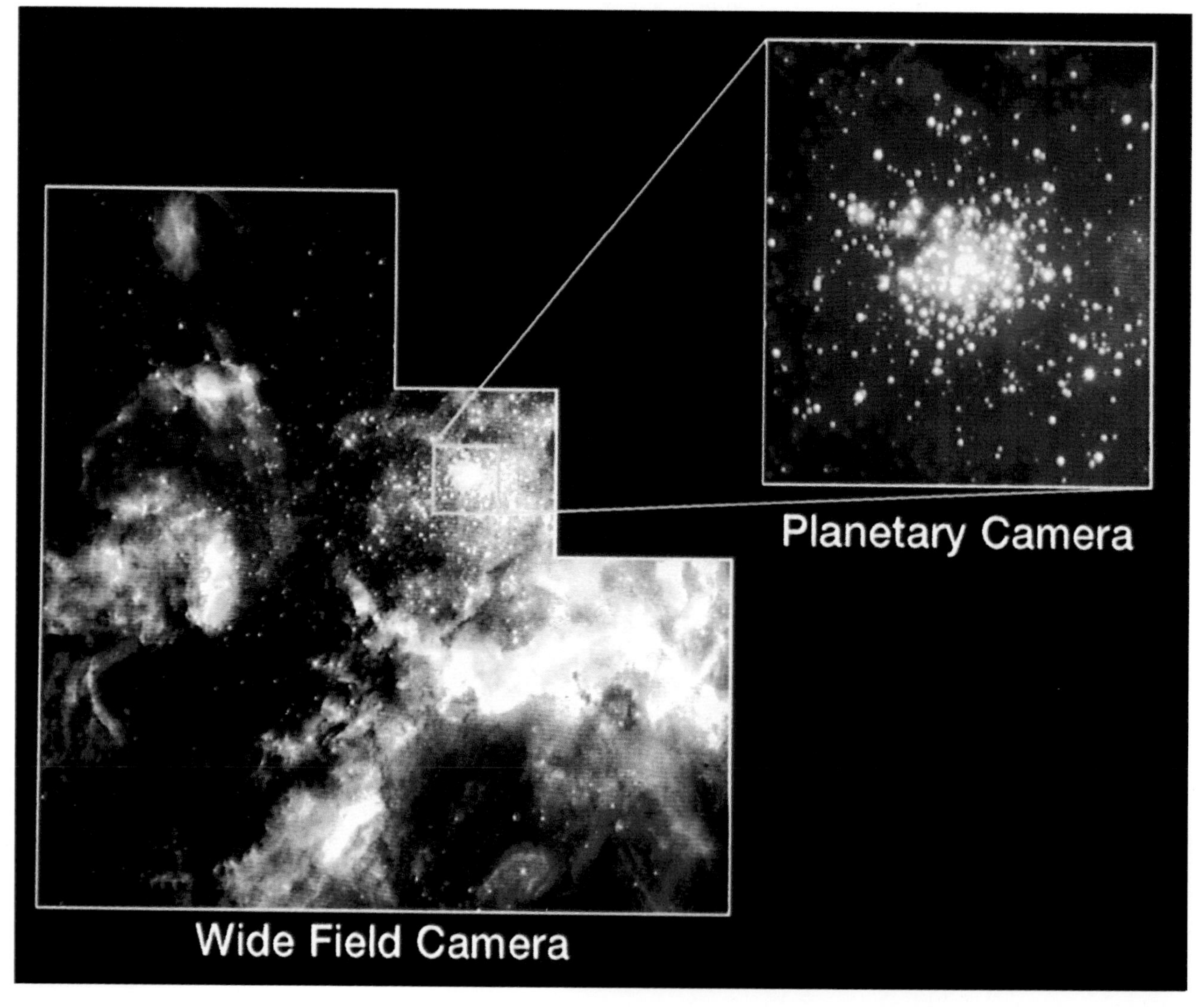

odds being against finding stars and planets, and then stellar systems with inhabited worlds inhabited with intelligent, cultural, technical, and scientifically capable life-forms. It is quite the other way around. The odds are trillions to one against the idea that we could never encounter world companions to our own planet, no matter how far removed across the vastness of the cosmos.

"They" are out there. One day, perhaps, we will meet.

Planet hunting has its own distinct and measurable rewards, but as NASA's Chief Scientist for the Hubble Project, Dr. Edward J. Weiler, explained, "The best things we expect Hubble to find are those that are completely unexpected. We're going to be surprised again and again. The universe is a treasure chest into which we've hardly yet started browsing."

No better example proved correct Dr. Weiler's predictions than when scientists aimed Hubble's cameras at a star-forming region of space—the 30 Doradus Nebula.

This star-birthing region extends across a vast expanse thousands of light-years wide, and contains in its center the extremely dense star cluster R136. At a distance of 166,000 light-years from Earth, R136 was a stellar whirlpool that caused great frustration for astronomers. For decades, using the most powerful ground-based telescopes, scientists believe R136 was a gathering of a small number of supermassive stars, immersed within the dust-and-gas obscuration of the dazzling 30 Doradus Nebula.

Hubble's dead aim to the heart of the cluster brought on scientists' mixed amazement and sounds of disbelief, followed by unrestrained jubilation.

The enormous clouds of glowing gas—ionized hydrogen lit up like masses of kerosene-burning lamps from the ultraviolet light emitted by extremely hot stars—had concealed a remarkable hotbed of stellar birth, a cosmic nursery completely unexpected. Even more to their surprise, scientists discovered this particular region was filled with hundreds of the brightest and most massive stars yet known anywhere in the universe. All this existed virtually at the doorstep of Earth, a paltry 166,000 light-years distant.

Paltry, certainly, when Hubble was targeting stars and star systems many billions of miles across space.

Focusing every more tightly, Hubble's technicians peered deeper through the gauzelike flowing nebula. Here stretched an area that had been believed to be mainly dust and gas, mundane in its appearance and what it might contain, a sort of gaseous backwater compared to other wonders abounding in almost every direction. Slowly but surely, as the cameras' resolution was tweaked

to near perfection, there emerged instead hundreds of dazzling great stars.

Now the Hubble team gazed in wonder at more than 3,000 massive young stars. The incubator was a ceaseless torrent of blinding light, soaring temperatures, and stars so mighty they made our own sun pale in comparison. From these pictures, aside from their magnificence and beauty, the Hubble team discovered the keys they feel will confirm how the stars throughout the Milky Way came to life. The nebula 30 Doradus lies just beyond the major rim of the Milky Way, but the halo of gas and dust and faint stars of our galaxy intermixes with the tenuous reaches of 30 Doradus, so that what is learned from one great cluster will apply equally to understanding of both.

Most impressive, even startling, stated the Hubble scientists, was the confirmation that stars much less dense than our sun actually burst into nuclear blaze with the same crushing rapidity and violence as the truly massive suns. For the first time, through Hubble, we now had direct photographic proof of immense starbursts, keys both to the past and to the future of our galaxy.

Left: This Hubble photograph shows rich detail in a pair of star clusters 166,000 light-years away in the Large Magellanic Cloud (LMC), in the southern constellation Doradus. Hundreds of dazzling great stars appeared where there had been—or it was believed there had been—a stellar marsh of little significance. *(Space Telescope Science Institute and NASA)*

Ten thousand years ago, Eta Carinae, 150 times more massive and more than four million times brighter than our sun, exploded so violently that its light, when it reached Earth after a hundred centuries, appeared in the heavens second only to Sirius—8.6 light-years away—as the brightest star in the night vaults. Eruption, detonation, shock waves; all were the signatures of that titanic stellar blast.

Then Hubble zeroed in on the remnants of the explosion that shattered nearby stars. The matter ejected by the stupendous blast is a series of glowing rings and halos of stunning rich red color, mixing with filaments of light blue and yellow and orange, all followed in a continuous, nonstop expansion of violence by white-blue nebulosity. Hubble also captured an image never seen before in the millions of pictures taken by astronomical observatories around Earth—two vast mushrooming lobes of stellar detritus, a massive dumbbell-shaped wreckage of an angry, outreaching explosion—each dimpled and rumpled misshapen lobe greater in size than our entire solar system from one edge of the Oort Cloud to the other. Eta Carinae is a marvel unto itself, but like a ship voyaging through exotic new lands, the cameras of Hubble swept onward to other kingdoms.

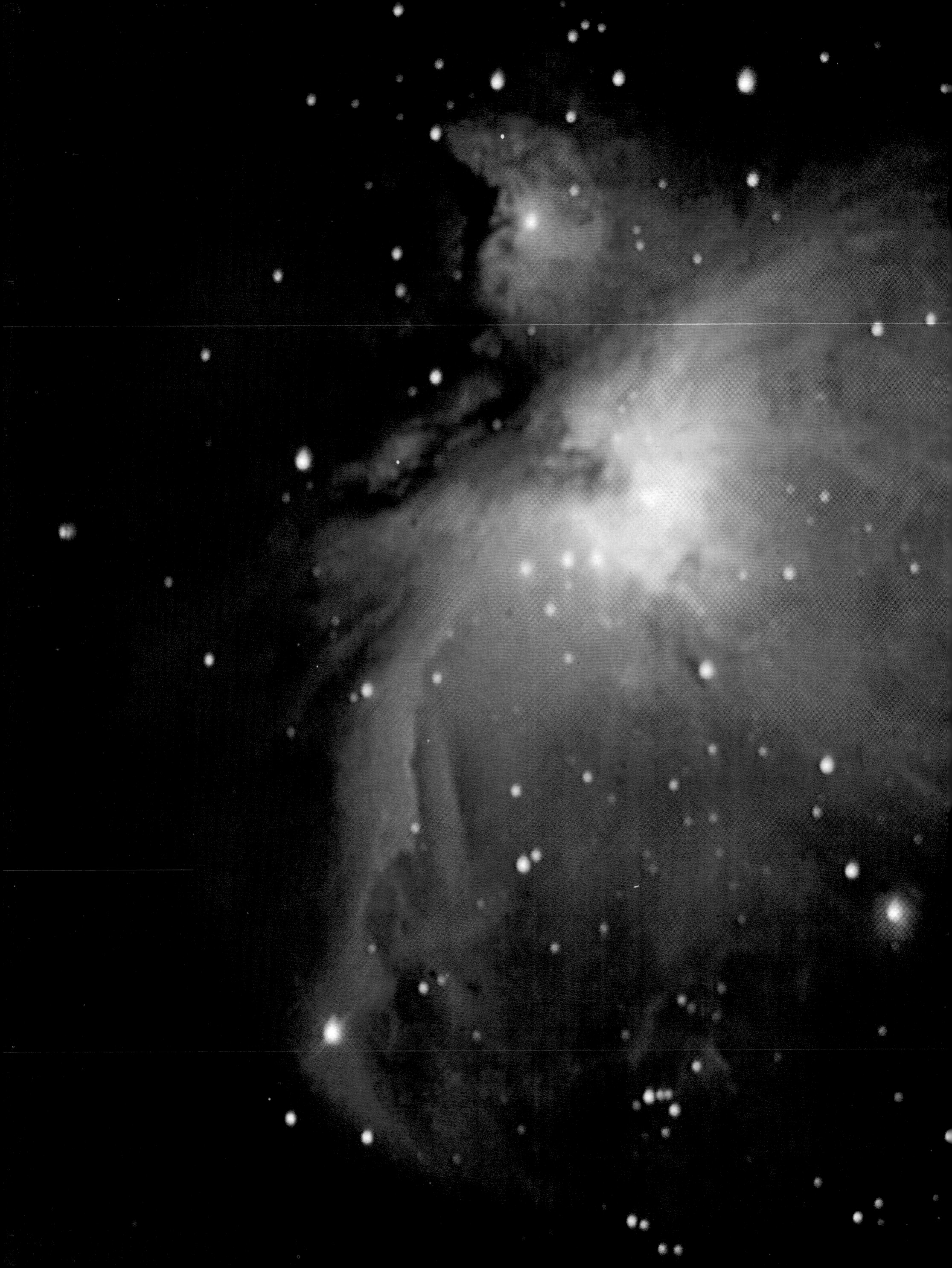

13 Magic of the Stellar Kingdom

Our galaxy rose first from legend of old, a glowing in the dark to hold off ultimate blackness when the moon was new and hiding from man. At those times only the bridge of faint lights traversing the heavens from earthly horizon to horizon held back the beasts of darkness eager to savage helpless humans. This was the dominant theme of the !Kung Bushmen of Botswana. Where they lived the band of stars in our Milky Way stretched directly above them, and with good reason they judged this glow as "the backbone of night." This is the time when great beasts sought their victims.

To the !Kung the light in the dark sky was the spine of some celestial animal. Were it not for this powerful arch, then the sky, burdened with blackness, would collapse from that weight, sending great blocks and shards of nightfall to destroy all living creatures on Earth.

Judging the world from the religion of the !Kung, this made eminently good sense. That bridge of stars, the skeletal band of light, was their protection against disaster. Cloudy night skies must have been sheer terror for these people when the "backbone

Previous page: In the enormous depths of space, there swirl vast clouds of cosmic dust and matter. These are known as nebulae. They are the birthplace of stars. Our own sun formed in just such a nebula. Here we see the overwhelming beauty of the Great Orion Nebula. *(U.S. Naval Observatory)*

Below: As seen in this photograph of the central Milky Way by the IRAS spacecraft, the light in the night sky appears like the spine of a celestial animal. *(NASA)*

of night" was lost completely to sight.

In years following, new cultures spoke of the celestial lights as the gardens and homes of the gods whose whims and fancies dictated everything that happened on the Earth far below. Natural disasters were actually whims and angers of the gods. Few had more power than Hera, goddess of the heavens and wife of Zeus. The rulers of Olympia, so went the legends, flew to Samos for their honeymoon. There milk issued from the breasts of Hera, spilling across the entire heavens, sparkling and gleaming. Thus was born the Milky Way as our galaxy, and so it has been from the time of the Greeks.

Had the ancients known of the brilliance and fantasies of shapes and color, the awesome power in the stars they never understood were suns, some larger or smaller, some yellow or blue-white or red, they might well have attributed these wonders as well to the talents of Olympia.

One magical look through the camera of Hubble, capturing the color-emblazoned heart of the Great Orion Nebula, would have

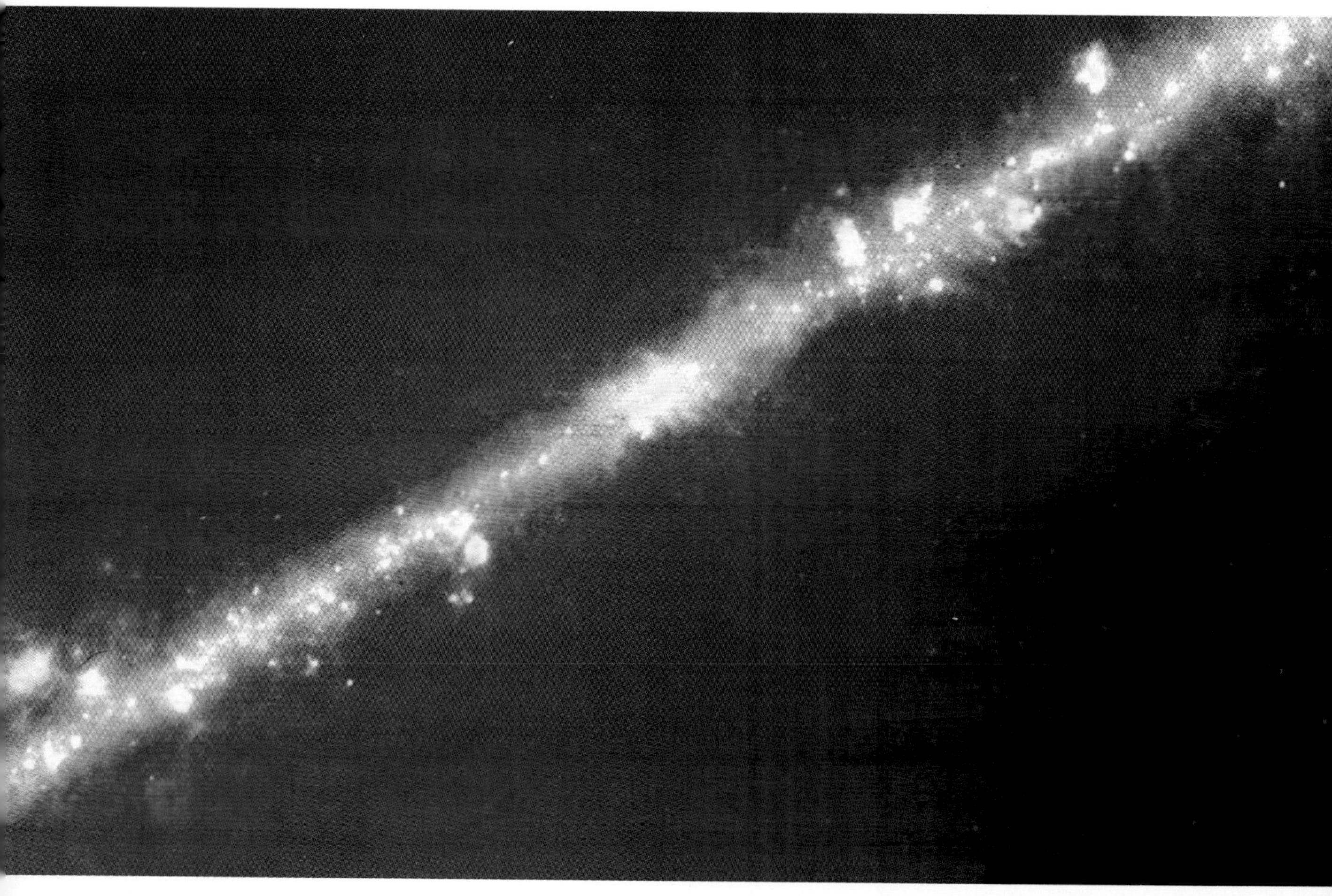

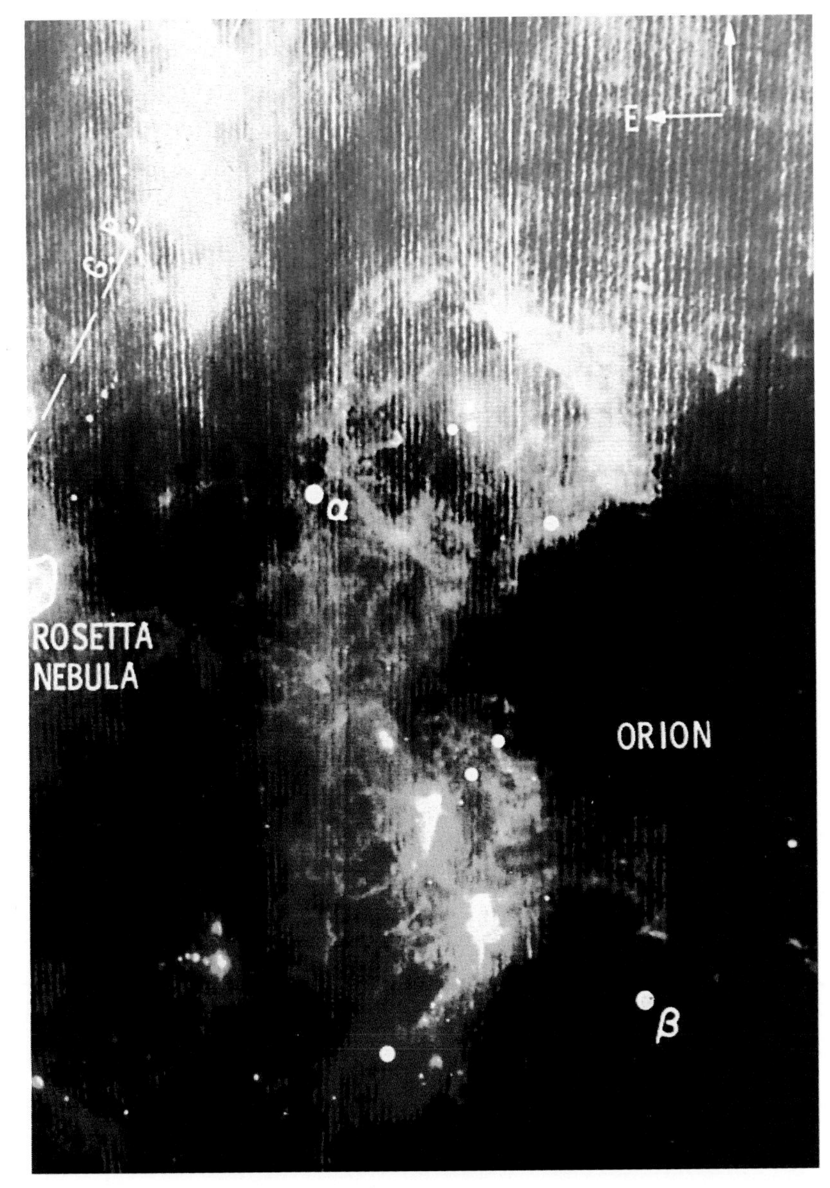

Left: The enormous gas clouds of Rosetta and Orion in the Great Orion Nebulae, as seen in this IRAS spacecraft photograph. *(NASA)*

left no question that only gods could dwell amid such splendor. The enormous gas cloud of Orion flares with bright and ghostly hues flung away from newly formed, violent young stars. Shock waves from distant exploded suns rush through the Orion gases to lash them into ever-changing forms and dramatic hues. This is the heart of star birth where Orion the Hunter gleams brilliantly in the night heavens. The disk of dust and gas stretches 20 billion miles from side to side, containing in its midst not only the powerful stars but also the spattering glow of millions of young stars. Through the splendor, enormous jets and plumes streak as dazzling spears of light, hurled forth from the births of young stars erupting into thermonuclear life.

There are strange beacons through the Milky Way, announcing their presence by waxing and waning in periods of intense light.

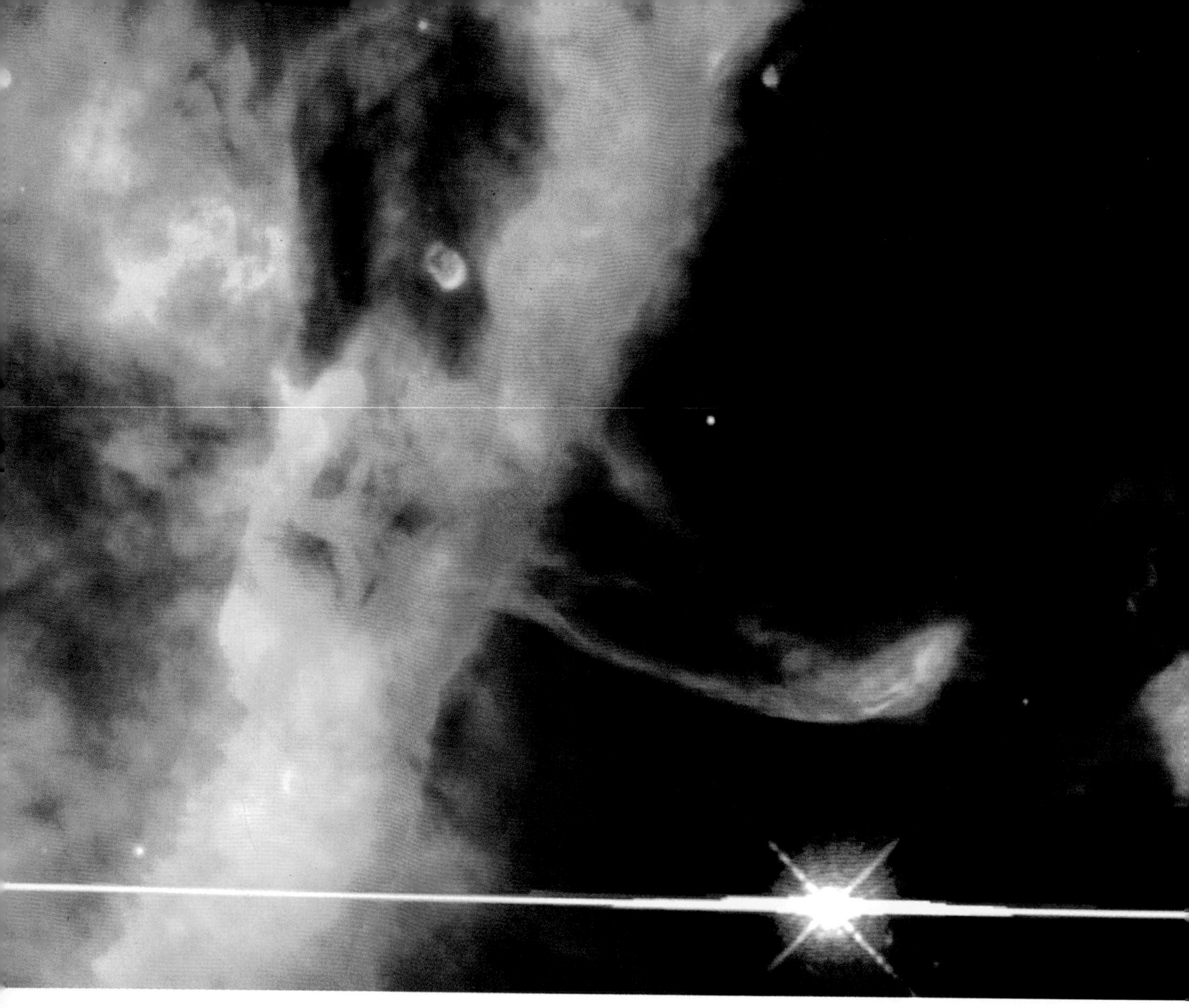

Above: The Orion Nebula is a rich spawning ground for new stars. *(Space Telescope Science Institute and NASA)*

A Hubble astronomer, viewing the astonishing panorama of the Cygnus Loop, the whirlwind remains of a cataclysmic supernova 15,000 years ago, remarked to reporters that "if the mythological gods must be living anywhere in outer space, then this fantastic, magical wonder of light and beauty is where they are to be found."

The remains of the Cygnus Loop, twisting and plunging through space at more than three million miles an hour, were captured finally in a stunning image by the Hubble telescope. It proved to be a photograph of what NASA's Dr. Edward J. Weiler described as "unprecedented clarity," as he studied the structure behind the shock wave.

The blast shock is described by the Hubble team review of the image as "slamming into tenuous clouds of interstellar gas." Ongoing for thousands of years, the collision compresses the gas until it breaks up its atomic structure and begins to glow.

Below: This is an image of a small portion of the Cygnus Loop, which is the whirlwind remains of a cataclysmic supernova 15,000 years ago. *(Space Telescope Science Institute and NASA)*

The bluish ribbon of light streaming across the glowing shock wave is a burst of stellar debris finally catching up with the shock wave, which has been slowed down by resistance from unusually dense clouds of debris from past explosions. Few scenes photographed by Hubble can match this kaleidoscope of colors. Oxygen atoms pulse blue light at 100,000°F, hydrogen atoms glow in green, and finally the red hue fans out from sulfur atoms from the gas medium cooling down to a "mere" 18,000°F.

Below: The magnificent Lagoon Nebula is found in the constellation of Sagittarius. *(U.S. Naval Observatory)*

When a star becomes a nova, the event is a titanic release of energy with no way to predict its final post-explosion form. The Crab Nebula, shaped astonishingly like jellyfish, is the stunning detritus of such a blast. In A.D. 1054, a supermassive star blew itself apart with such light intensity that it flashed in the heavens with a pinpoint glare that could be seen in broad daylight. Chinese astronomers recorded the event in detail, and there are indications that the star also troubled and fascinated Native Americans in what is now the southwestern United States. For several months it dominated stellar light by dark and, amazingly, continued to gleam in daylight.

Nine hundred years after it exploded, remains of that star still expand outward, churning and twisting slowly through the crab-like shape; its edges and streamers seem to be alive.

The nebula glows so brightly because the great star from which it was created now exists as a pulsar—a pulsating star of immense energy. A superheavy star that collapses in such a manner squeezes down to a mere ten miles in diameter. The energy source of the Crab Nebula blasts all matter within its reach with a colossal magnetic field and crushing gravity. The nebula will continue to expand, changing color and shape, for many thousands of years. Finally it will collide with other star debris, and from the roiling collision matter will coalesce into clumps that will follow the long road to creation of protostars and then brilliant suns.

Nebulae are sprinkled across our galaxy and are scanned in every direction by the Hubble Telescope. They seem to have been created in all shapes, sizes, and colors, and each seems also to contest the others for stunning beauty.

One of the most impressive is the magnificent Lagoon Nebula in the constellation of Sagittarius. It is a cloud vast and filmy, the outer mists behind which thousands of new stars are coming to life. Light from the energetic young stars imbues the gaseous film with a striking red color, but slowly that hue will change as gaseous energy wanes and the expanding clouds mix with other, older nebulae refuse too dim even to be photographed by Hubble. When future telescopes train their cameras on what is now the rosy hue of the Lagoon Nebula, the color will have shifted to a bluish tint, the billowing space cloud painted anew by the energy of a huge blue-white star already being formed behind the expanding pulse-wave.

The Aquarius constellation presents the Helix Nebula (NGC 7293) as a look into the far future of our solar system. A glowing ring billows outward from a supernova explosion, so light and filmy in appearance it seems to be ghostly fire making up its planetary nebula shell. One day in a future far from now, when our sun declines in its stellar old age, it will likely enlarge through the red giant stage, then blow off its outer gases to create what is called a "planetary nebula shell" as premiered for us by the Helix.

The nebulae of the Milky Way are a preview of what astronomical searches have already found in distant galaxies—the common threads left by exploded stars and powerful shock waves racing through the interstellar medium. The U.S. Naval Observatory captured the Veil Nebula, a lacy, almost phantomlike flowing

Above: The Veil Nebula. *(U.S. Naval Observatory)*

of interstellar gas clouds illuminated by stars behind the expanding space mists.

Rare among the nebulae of soft gaseous shells is the Dumbbell Nebula of Vulpecula, another image captured by the U.S. Naval Observatory—and a type of nebula slated for search by Hubble.

Of all the nebulae discovered, nothing compares to Dumbbell, easily the most prominent nebula found in the sweeping search for stellar phenomena. When first discovered, features of this formation were identified as planetary nebulae due to their disklike outlines, often mistaken as planetary formations. They are huge expansions of exploded stars that blew out much of their mass equally on two sides, and eventually will disperse into softly glowing gaseous envelopes, sooner or later to be shocked by the blast waves of other exploded stars into starting the birth of new stars—the reincarnation effect of the cosmos.

Until the repaired Hubble Space Telescope drifted away from its repair shuttle in December 1993, astronomers endured a dichotomy of emotions regarding the photographic images sent down to Earth from orbit. Images sent from the flawed Hubble were less than hoped for, yet they were still superior to anything from all other cosmic photography.

Hubble had aimed its optics at a violently ballooning "bubble" of gaseous clouds blasted from the surface of Cygni 1992, an explosion first detected when light from the stellar eruption reached

Earth on February 19, 1992, after a voyage of 10,430 miles at the speed of light. Photographs from ground-based telescopes under the best possible conditions offered scientists little better than a fuzzy dot in a suspected maelstrom of expanding, swirling gas.

Though the first image from Hubble was far from what had been promised, it was a splendid improvement in resolution. Out of an image of fiery masses and surrounding glowing gas clouds there was discernible a mysterious barlike structure running through the aftermath of the exploded star.

After Hubble went through servicing, it was programmed to take more images of Cygni 1992. These photographs stunned astronomers with clear resolution of details they had never expected to find. Seven months had gone by since Hubble's previous photographs of Cygni; the COSTAR photographic systems captured events rarely photographed—major changes in the violent move-

Below: The Dumbbell Nebula of Vulpecula. *(U.S. Naval Observatory)*

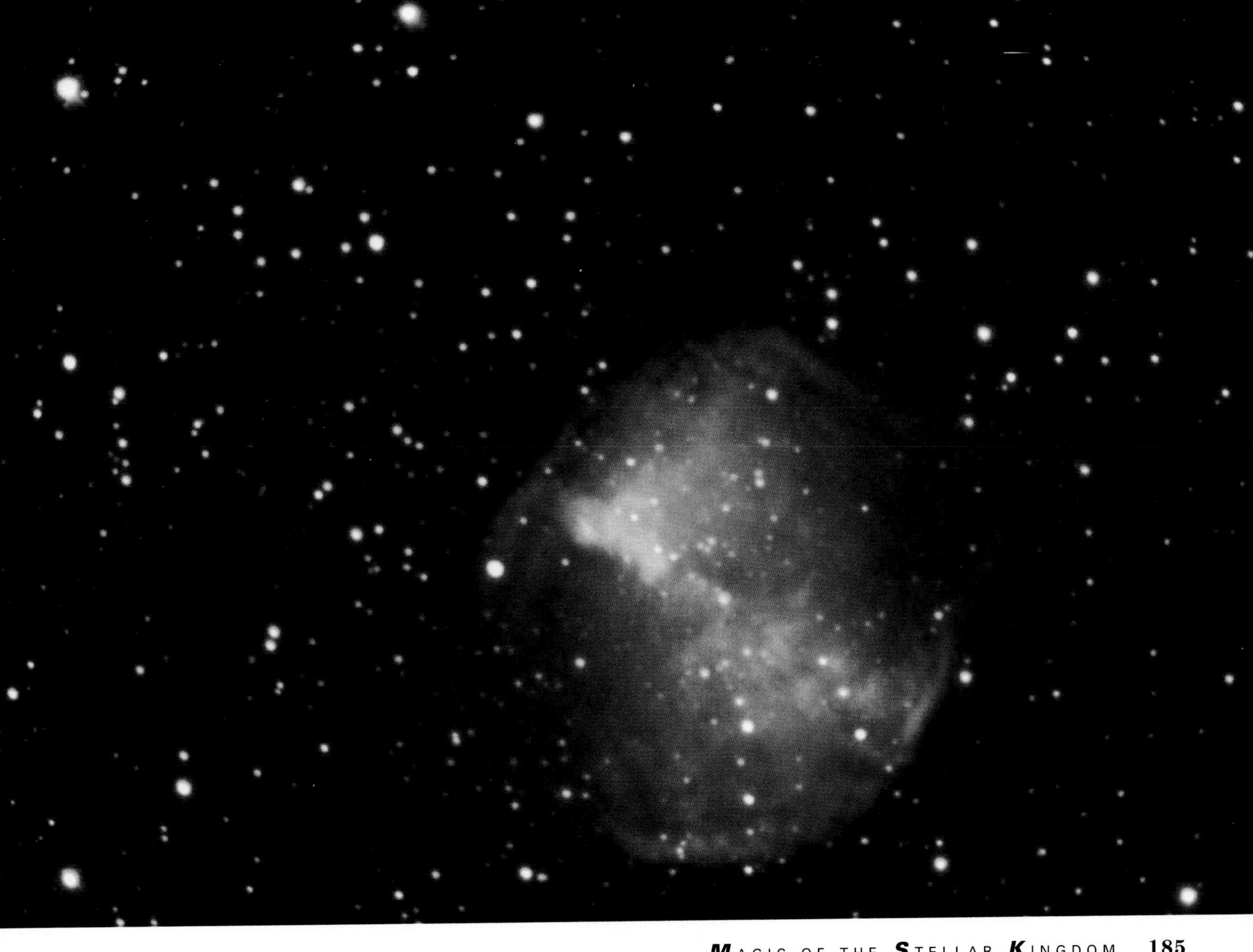

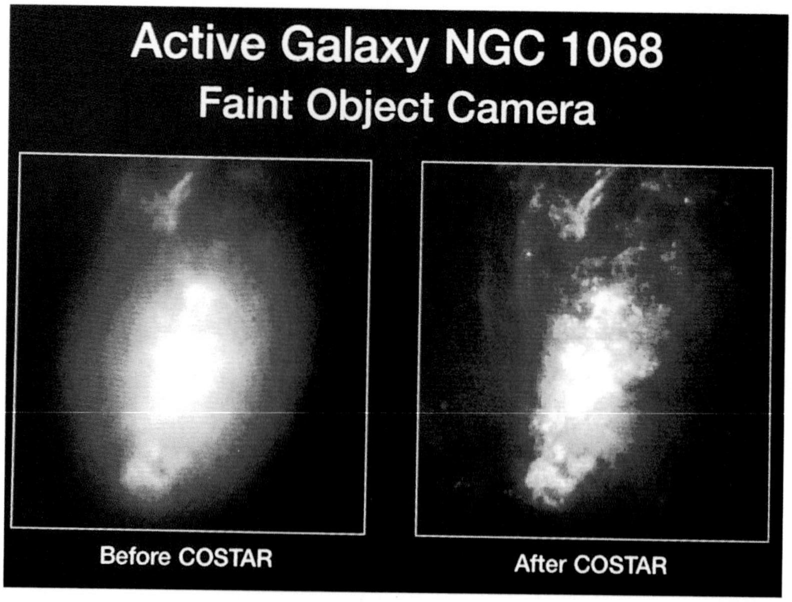

Active Galaxy NGC 1068
Faint Object Camera

Before COSTAR

After COSTAR

Right: In the photograph on the left—pre-COSTAR—Hubble took this picture of a gaseous explosion from the surface of Cygni 1992.

Hubble went through servicing, updating of equipment, and correction of its seeing ability in December 1993, and this next image of Cygni 1992—with COSTAR (on the right)—stunned astronomers with clear resolution of details they had never expected to find. *(Space Telescope Science Institute and NASA)*

ment of the exploded star. In those seven months, the distorted ringlike structure had expanded from 74 billion to 96 billion miles. The curious barlike structure had vanished completely, which Hubble team scientists interpreted as a surprisingly swift dissipation of the gaseous debris. Deep within the enormous sphere of glowing gas there remained a white dwarf star, a compacted remnant of a once-great sun.

The Hubble Space Telescope is the crown jewel in a vast array of instruments that survey our universe. Ground-based telescopes stationed about the world, giant radio telescopes to capture energies invisible to our eyes, and satellites far from our atmosphere to scan the far heavens in ultraviolet and infrared light, as well as capturing a wide spectrum of nuclear radiations—they all combine to complement one another in charting the universe and its near and distant objects.

When Dr. Edward J. Weiler announced the greatest scenes from space would be of discoveries we could not even hope to anticipate, he was dead-on.

But of all the "unexpected discoveries" so far revealed for the first time, none can compare to finding an entire galaxy of more than 100 billion stars in what NASA astronomers described as "a galaxy in our own cosmic backyard." The unprecedented discovery of "the newly discovered city . . . of stars" was an effort complimentary to Hubble, which only now will be brought into the research effort to direct its most powerful cameras at what shone brilliantly in space—but hidden from view by the mass of our own Milky Way.

It is a splendid affirmation of international teamwork supporting the shared goal of exploration.

A research team in the Netherlands had for years worked with the Dwingeloo Obscured Galaxy Survey (DOGS) radio telescope at Dwingeloo to detect radio emissions from galaxies that might be obscured from the view of optical telescopes, such as Hubble, by the Milky Way. On August 4, 1994, DOGS—the accepted acronym for the long-term search—was well into its ongoing "blind search," hoping to pick up any stray signals that might reveal otherwise invisible and unknown radio-emission sources. The scientists were alerted that something was coming in. The radio telescope had made preliminary identification of a radio signature. Soon they knew they'd snared their quarry—still unknown and invisible but betrayed by its radio wave emissions.

Then they knew they had bagged a real prize; the radio signature matched such signals received from known spiral galaxies. The "image" detected was about half the size of the full moon as observed from Earth.

As fast as they assembled their data, the DOGS team contacted their colleagues manning telescopes on La Palma in the Canary Islands, in Hawaii, and in Israel. Those teams focused tightly on the source of emissions as reported from DOGS.

The secondary observations confirmed their wildest hopes—they had detected a barred-spiral galaxy of more than 100 million stars only 10 million light-years from the Milky Way. The telescopes about the world confirmed in visible and infrared wavelengths that not only was there this enormous spiral galaxy, but that a nearby group of galaxies, also heretofore unknown, was also obscured by the Milky Way. These were named Maffei 1 and 2, and are believed to be only the first additional galaxies of a large group that have affected the Milky Way through their gravitational force, unknown and unsolved, for reasons now obvious.

"For a long time, astronomers did not realize the importance of interstellar dust in shaping our view of the universe, and simply ignored parts of the sky close to the plane of the Milky Way when it came to studying extragalactic objects," stated Dr. Harry Ferguson of the Space Telescope Science Institute in Maryland. "It is sobering to realize that a galaxy as close as Dwingeloo 1 could have remained hidden until now, and could have a real impact on the understanding of the motion of the Milky Way through the cosmos."

And Dwingeloo is just the beginning of Beyond the Milky Way.

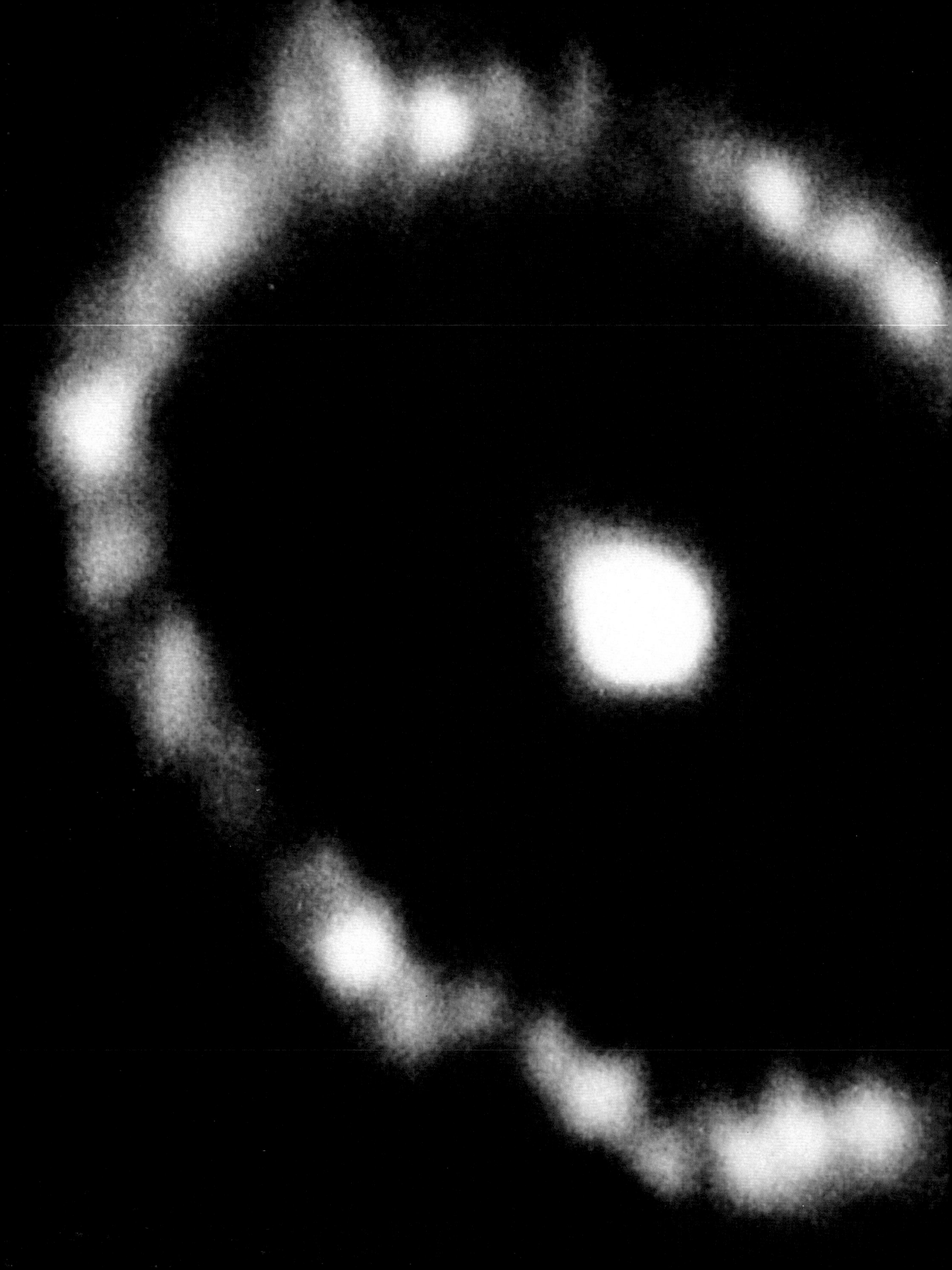

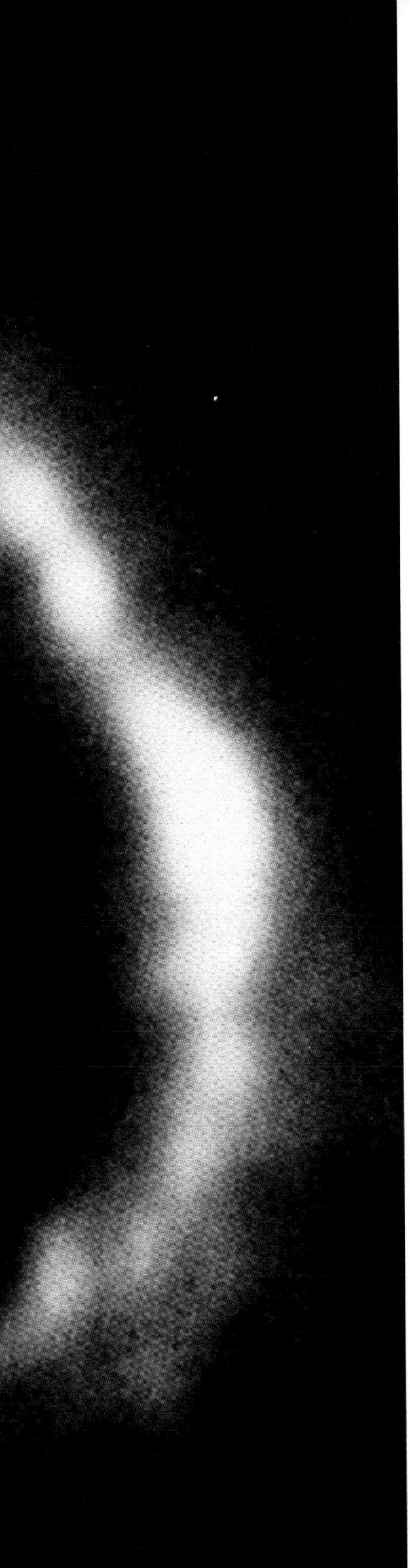

14 The Outer Limits—Or Are They?

The universe is filled with galaxies of startling shapes, sizes, colors, speeds. Many are new, others so aged they have dissipated into stars dimmed by lifetimes lasting billions of years. Science once accepted the Milky Way as the only galaxy in a universe that was no older than one billion or two billion years.

The renaissance of galactic knowledge came slowly, as hand-ground telescopes pushed back the limits. Precision-made telescopes mushroomed from a few inches to 200 inches across their light-gathering mirrors and, almost abruptly it seemed, we looked at galaxies by the thousands, then the millions, on up into many billions, and now no number of galaxies can be counted with even a glimmer of accuracy.

Peering deeper and deeper into the void, looking back ever further in time, our awareness of the real universe mushroomed. From great galaxies we discovered there wheeled small groups of galaxies, then clusters, and superclusters of galactic islands spanning millions of light-years. The count of all these great cities of galaxies confirms that stars number in the high trillions, likely more than we will ever know.

Previous page: This is a Hubble view of a single ring of the Supernova 1987A, which was first detected in Earth's southern hemisphere after its blast in the Large Magellanic Cloud Dwarf Galaxy. *(Space Telescope Science Institute and NASA)*

Below: Hubble found these "Peculiar Galaxies" struggling to develop from their mushroom stage into galactic islands of wheeling stars, spanning hundreds, thousands, even millions of light-years across. *(Space Telescope Science Institute and NASA)*

The more we learn of this kaleidoscopic universe within reach of instruments and optics, the more puzzled we become at long-standing definite conclusions dissipated with newfound mysteries, and paradoxes that seem as common as the stars themselves. We are caught in that age-old web of the more we know, the more we realize how little we really know.

But it is all wonder, power beyond comprehension, and a variety of stellar creations we have just begun to count. Each year—now each month or less—Hubble's powerful new eyes upon the universe discover stars and energies that defy the hard logic on which we have relied for so long in our quest for understanding.

For example, a gamma-ray burst was detected in 1979 from the direction of the Large Magellanic Cloud, our nearest galactic neighbor. In one-tenth of one second the gamma-ray blast unleashed more energy than our sun will generate in the next 10,000 years.

One would expect such violence to be an immense ragged eruption of flames and smoke, punctuated with debris and blazing

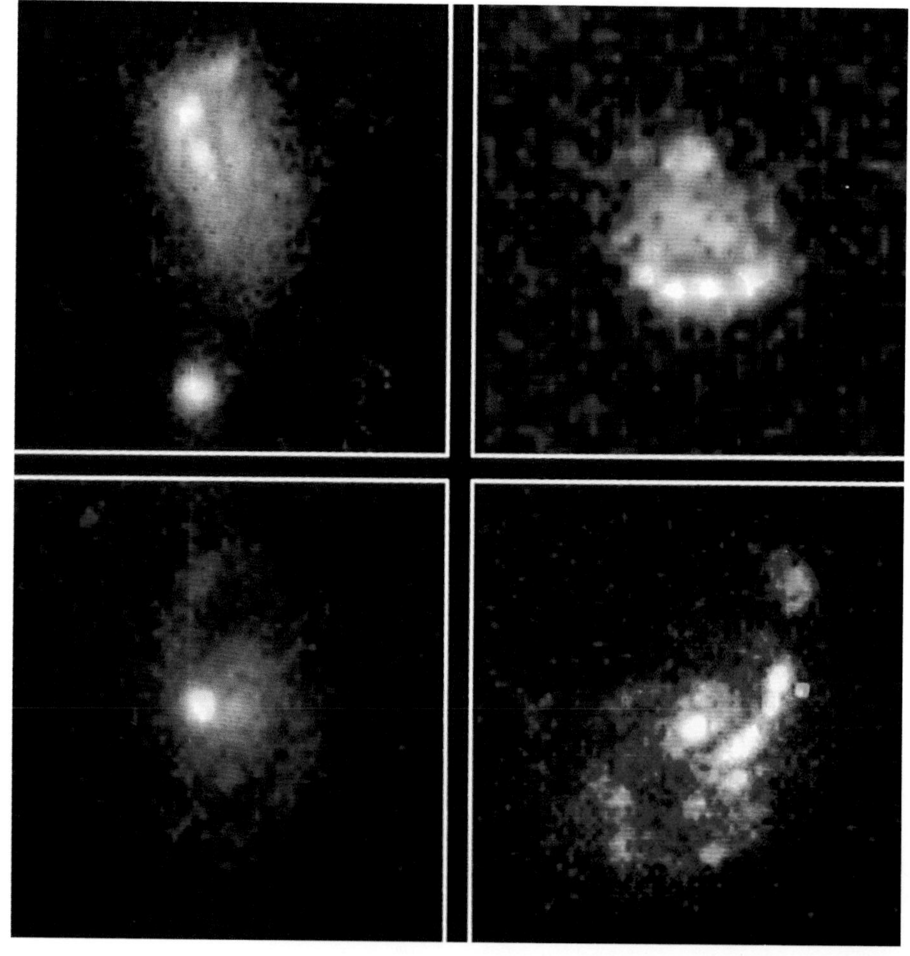

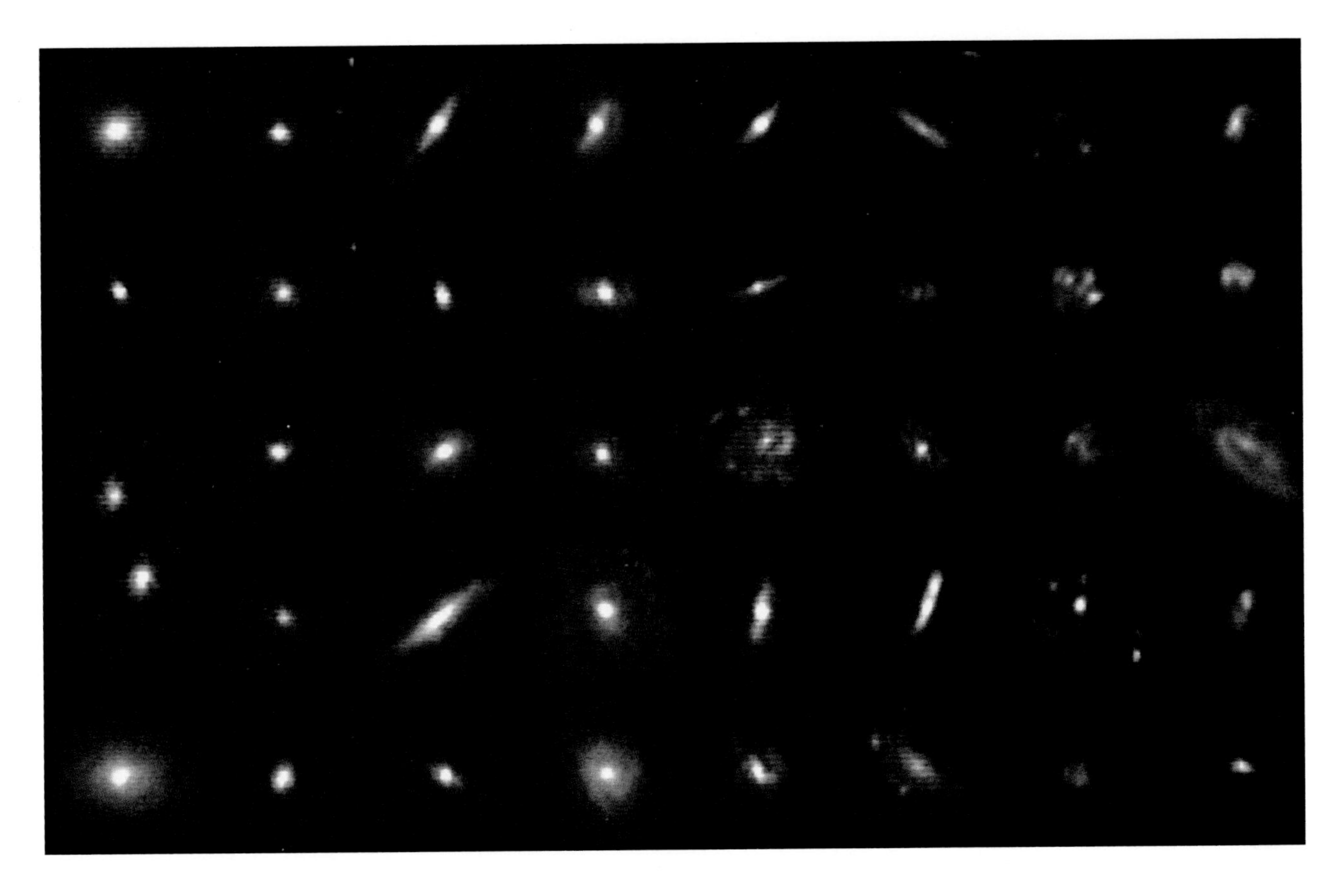

Above: Forty "cities of stars" are seen here in this Atlas of Galaxies by Hubble. *(Space Telescope Science Institute and NASA)*

chunks of a star. The last thing to expect is a stellar-sized gamma-ray burst in the form of a beautiful flower on a stem seemingly arising from a void. That flower is a single blast of energy exceeding the output of millions of stars hurling forth gamma rays that hurtle through physical matter as if it were so much smoke.

Exceeding even the unexpected nature of this massive gamma-ray eruption is this scene (see photo, page 192) of such an event about to happen. It is a case of cosmic billiards on a scale of energy against which the term *vast* fades as an understatement.

Two neutron stars have somehow fallen into tight orbits around each other. Each star was once a massive supergiant that went supernova. In the compression implosion, the nuclear structure of each star was rammed down to solid spheres of pure neutrons. Each is unimaginably dense, where an object the size of a child's marble weighs billions of tons. Were such a tiny sphere dropped from just several feet above the ground on Earth, in a flash it would penetrate the entire 8,000-mile diameter of the planet and hurtle on past at thousands of miles a second.

Ultimately, in so tight a pair of frenzied orbits, the surfaces of both stars will touch. It is a stupendous clash of energy. The slightest contact between two neutron stars is an instant flash of energy so dazzling its light washes out the sight of everything for tens of millions of miles, including brightly burning stars.

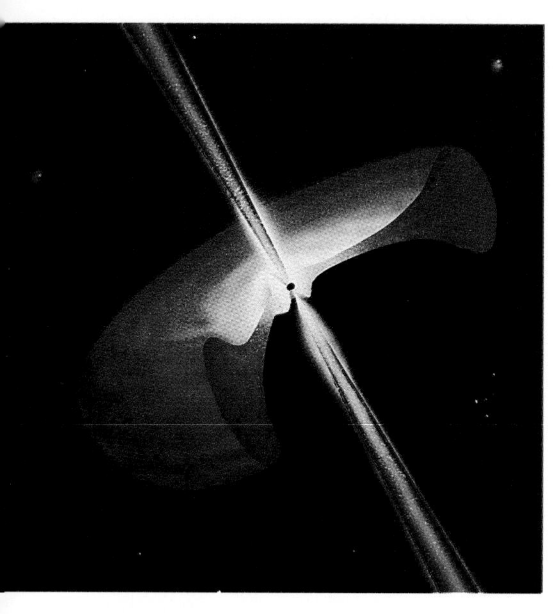

Above: Gamma-ray bursts generally occur in the far reaches of the observable universe. They are the most energetic, powerful objects we have ever seen. This NASA photograph shows a gamma-ray burst releasing in one-tenth of one second more energy than our sun will generate in the next 10,000 years. *(NASA)*

At such a moment, the contact between the two neutron stars becomes the single most energetic and powerful object ever seen, known of, or postulated anywhere in the universe.

Galactic structures form major targets of Hubble, but not simply the entire island universes. Hubble has that superb ability to zero in to specific objects, seeing with a clarity never expected even after the first repair mission. Photographs of Spiral Galaxy M-100—tens of millions of light-years from Earth—were captured with so fine a resolution that individual stars were clearly defined from the overall galactic structure. A single image from Hubble made from three wide-field cameras and one planetary camera performed additional image legerdemain, depicting almost the entire galaxy, itself only one part of the huge Virgo Cluster.

Other Hubble images that excited astronomers were views of M-31, the Andromeda Galaxy. Close-up images of the M-31 core revealed a rare scene of a double-nucleus galaxy; the brighter of the two cores was suspected to be a torn and cannibalized core of another galaxy drawn into the main body of M-31 and stripped of its outer formations of stars.

Another example of stars stripped of their gaseous envelopes was found in the Hubble images of the core of Galaxy M-15, where astronomers found 15 violent blue stars, stripped of their outer envelopes by their own extremely hot cores.

Few images proved as exciting to astronomers as those of the Gravitational Lens G2237-0305, better known by the name Einstein Cross. The one light source of the quasar shows up in images as five separate superpowerful stars.

RINGS AROUND THE NOVA

Exploding stars, jets of glowing matter, rings about planets, light bending as if it were fiber optics, coruscating nebulae—they're all now fairly commonplace wonders of our universe.

In May 1994, Hubble captured a sight so bizzare and unknown that at first the science teams were certain a mistake had been made with the photographic imaging process of the space telescope.

Mysterious giant rings, wide glowing red loops intertwining with a massive core ring made of layers of bright red and gold, orbited the remnants of a shattered massive star. The rings rotated and wobbled in one of the most dazzling space shows ever seen—especially with rings on a stellar stage that's 1.7 trillion miles wide.

Above: Spiral Galaxy M-100—part of the Virgo Cluster. *(Space Telescope Science Institute and NASA)*

The mysterious mirror-imaged pair of rings of glowing gas encircle the site of the stellar explosion supernova 1987A. One hundred and seventy thousand light-years from Earth, a blue giant star obliterated itself. Hubble was directed at the enormous light flash remnants. The images caught scientists completely by surprise. They stared at strange and delicate rings of light suspended in space about where the star had been.

"It's beautiful! I even have it on a T-shirt!" exclaimed Richard McCray of the Joint Institute for Laboratory Physics. When asked how the rings might have been created, he shook his head. "I'm stumped. There's nothing else like it in the sky," he replied.

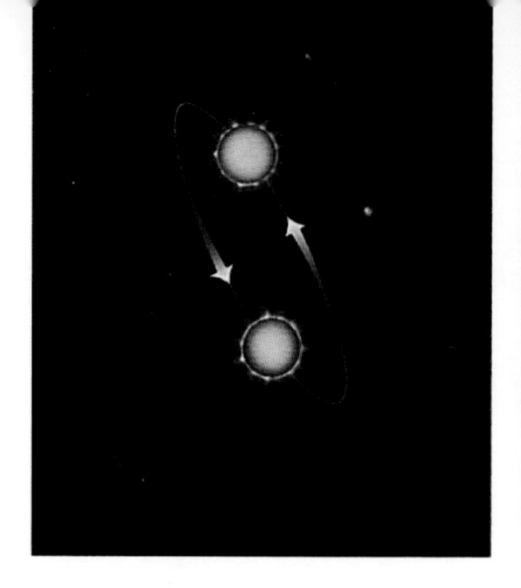

Right: Here in this NASA photo, on the left we see two neutron stars in tight orbits around each other. On the right, the surfaces of both stars inevitably touch, releasing a stupendous amount of energy. *(NASA)*

Astronomers throughout the world echoed McCray's reaction. Chris Burrows of the European Space Agency and the Hubble Team in Maryland said: "This is an unprecedented and bizarre object. We have never seen anything behave like this before."

Burrows offered one of several working possibilities. The rings—to the surprise of every scientist involved—proved to be part of an hourglass-shaped bubble of gas that completely encompasses the remains of the exploded blue giant star. They *could* be, in one scenario, remnants of a star that was a companion to the one that exploded. The vast rings could then be "painted" by an extremely high energy radiation beam, explained a NASA scientist, "like a spinning light-show laser beam tracing circles on a screen."

Supernova 1987A was first detected in Earth's southern hemisphere after its blast in the Large Magellanic Cloud Dwarf Galaxy. The event was immediately hailed as a bonanza for astronomical research, since it is the closet supernova to Earth detected in 400 years and the very first stellar violence of this sort ever to be studied by modern telescopes and other instruments. One member of the Hubble team hailed the event "a supernova in a test tube."

Astronomers considered that the glowing superloops of space may have spun off from the supernova 20,000 to 30,000 years before the star exploded. They were traveling that long in space when the supernova blast shattered the star 169,000 years ago. That light reached Earth in 1987, and on its way here suddenly illuminated the superloops so brightly they became visible and detailed to a telescope as powerful as Hubble.

But the best may yet to be—and not very far into the future. Shattered and ripped apart, Supernova 1987A still has plenty of violence as an afterthought. Stellar debris and fierce shock waves are still roaring through space toward Earth. Sometime about the end of this century these forces will smash into the inner ring of the exploded star—and that, predict scientists, especially with the most powerful telescopes in space and on the ground aimed at the coming stellar fireworks, will create three-dimensional scenes of nature in a fierce energy storm that can be captured and computer-processed into motion-picture film.

Above: Views of M-31, the Andromeda Galaxy. The photo on the left is a ground view of the galaxy. The center picture, also taken by ground telescope, shows the galaxy's core. But the close-up view on the right taken by Hubble shows a rare scene of a double-nucleus galaxy in the M-31 core.

Left: As seen in the single M-31 core close-up taken by Hubble, the brighter of the two cores was suspected to have been ripped from another galaxy and drawn into the main body of M-31, stripped of its outer formations of stars. *(Space Telescope Science Institute and NASA)*

STARS AND ALMOST-STARS

There are stars much smaller than our sun, others many times larger and thousands of times denser and brighter. Our sun is a pale yellow color in a panoply of stellar hues of white, blue, blue-white, red, brown, orange, and mixtures of these. There are white, yellow, and red supergiants, blue, white, and yellow, and pale red stars as main sequence suns, brown stars, dark stars, white dwarfs, red dwarfs, and pulsars—and unknown variations on the theme of black holes, which may no longer even be what we call a star.

Above: This Hubble Space Telescope image of the center of globular star cluster M-15. Stellar cannibalism could only take place where stars are so crowded together that chances for close encounters are exceptionally high.

The blue starfield is about two light-years across and is located 30,000 light-years from Earth in the constellation Pegasus. It is visible to the naked eye as a hazy spot about a third the diameter of a full moon. *(Space Telescope Science Institute and NASA)*

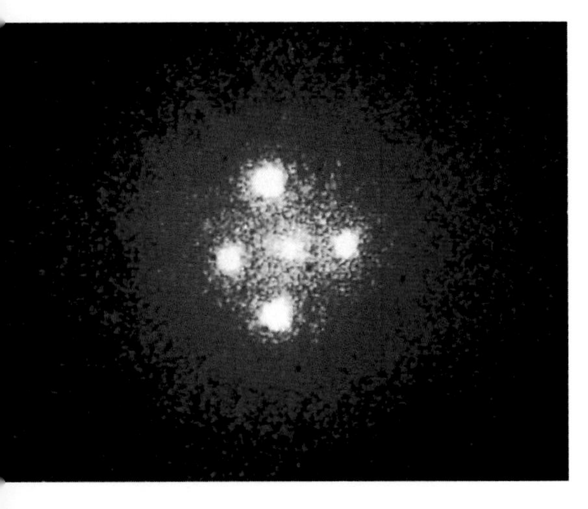

Above: A photograph made by the Gravitational Lens G2237-0305, better known as the Einstein Cross. Here the one light source of a quasar shows up as five separate superpowerful stars. *(Space Telescope Science Institute and NASA)*

A blue supergiant can be a hundred times the size of our sun, destroying itself with its voracious consuming of its own mass. So rapid is this burning and loss of mass that in several million years, it swells to several thousand times the size of our sun. Then, quivering like a deflating balloon, it accelerates its collapse until in an instant it explodes and then implodes, the latter process collapsing the stellar monster into a black hole that vanishes from our visible universe.

At the end of another branch of stars lie the red dwarfs that outlive all other suns. These quiet, steadily burning members of the star family are about the size of Jupiter, 70,000 to 100,000 miles in diameter. Their small size and lesser mass reduces the usual compression of the giant stars, giving the red dwarfs potential lifetimes of several hundred billion years. In stardom, bigger is better for only short periods of time. It's the quiet, small, dull stars that remain virtually forever.

Certain stars perform spectacular pyrotechnics. These are the cosmic blowtorches, energetic rivers of superheated gas spawned by violently burning stars. The "rivers" are electrically charged, many at least a million miles in length, long spears and lances of glowing flame. Sooner or later, as they race through the gravity fields of other stars, they twist and distort into strange curving shapes, bows, and twisted knots. Finally they either become swallowed whole by stars or are smashed into free-scattering stellar debris, awaiting their turn to be reborn into the stuff from which protostars form.

And there are the pulsars, which we have visited briefly. Until the advent of Hubble, pulsating stars were believed to exist only in the most distant reaches of space, many billions of light-years from our solar system.

Then in mid-1994, Hubble captured images of a new quasar, not many billions of light-years distant but only 600 million light-years from Earth in Cygnus A.

Dr. Anne Kinney of the Space Telescope Science Institute in Maryland, the main scientific center for Hubble, led the team that found this quasar. "I was stunned," she explained, "when we realized we had a quasar; it was a total surprise." Until Hubble imaged this "backyard quasar," all known quasars existed only in the regions containing the earliest times of the universe. "So it is unusual to find one in our own epoch," Dr. Kinney added.

The discovery of the quasar in Cygnus A provides astronomers with their first-ever opportunity to conduct detailed studies of the rare quasar. To a ground-based telescope, the mighty quasar hardly differs from the pinpoint of light of an average star.

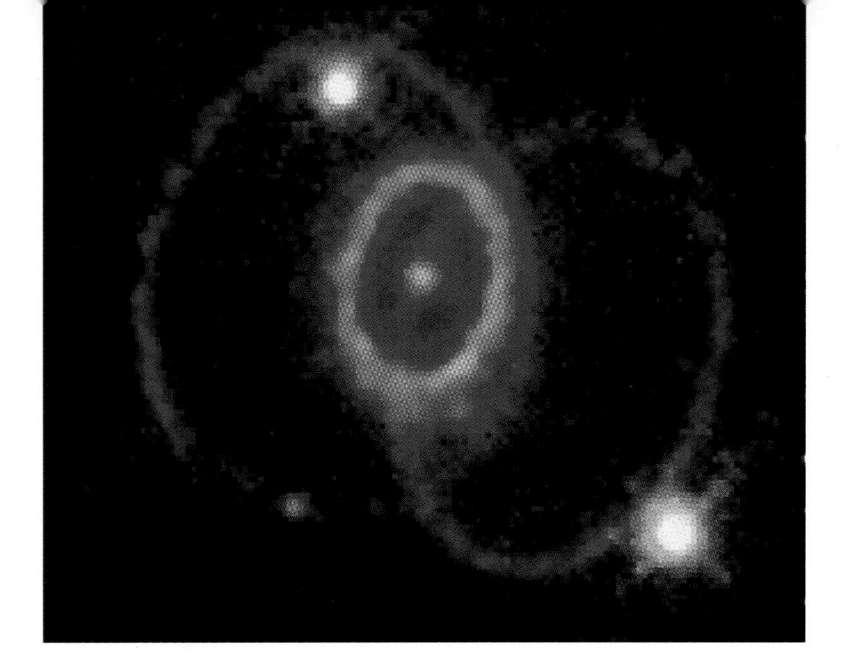

Right: This striking Hubble picture shows three rings of glowing gas encircling the site of supernova 1987A, a blue giant star that exploded 170,000 light-years from Earth.

Though all of the rings appear inclined to our view (so that they appear to intersect), they are probably in three different planes. The small bright ring lies in a plane containing the supernova, the two larger rings lie in front and behind it. *(Space Telescope Science Institute and NASA)*

It is anything but average—one quasar hurls forth hundreds of times more energy than an entire galaxy with more than 100 billion stars.

The quasars have an added imperative for detailed study. Quasars located in the remote galaxies appear to be powered by supermassive black holes that cannibalize dust, gas, and entire groups of stars to produce the vast energy fields they hurl through space.

Dr. Kinney emphasized in her report that this unexpected Hubble find implies that all radio galaxies might well harbor quasars still concealed from view by the very mass of material they eject.

Hubble first detected the Cygnus A quasar with the Faint Object Spectrograph of the telescope. Another instrument aboard Hubble—the Goddard High Resolution Spectrograph—captured the opposite end of the stellar scale: the birthing process of Beta Pictoris, an almost-star gas disk. Three major components of the rare formation were defined by Hubble; scientists categorize Beta Pictoris as a near-stellar region roughly the size of our inner solar system (the sun outward to the orbit of Mars). Since it cannot be imaged clearly in visible light, NASA produced an artist's illustration that shows a reddish outer ring, a diffuse gas disk which is already in stable orbit about the protostar. Within this outer ring is an inner disk which slowly is falling into the core of the growing protostar. Finally, closer in, are white cometlike curved streams within the bluish disk; these are dense steams of gas spiraling down the growing star's gravitational well.

Beta Pictoris remains high on the priority list of continued observations; Hubble will be ready to capture the moment of birth as nuclear fusion converts the steadily increasing mass to the blazing sphere of a new, complete star.

DARK MATTER

Cosmologists, those scientists whose interests lie all through the universe, have long held as inviolable that all space must be filled with a mysterious "dark matter." Like the ether of ancient astronomy, the dark matter cannot be seen, does not reflect light, and emits no radiation or energy of any kind that can be measured.

Yet it is the glue that keeps the universe from flying apart in a crescendo of disaster. Something must comprise at least 90 percent of all the mass in the universe. If the dark matter didn't exist, then rapidly rotating stars and galaxies would inevitably sling away their own substance in a frenzy of self-destruction. *Something* must be holding them together and balancing out the unequal forces of gravity and electromagnetic radiation.

Yet when asked what *is* dark matter, astronomers could do little more than shrug. Some theorized the universe was chock full of red dwarf stars, or dull brown stars, in such vast numbers that their combined gravitational effect served as a cosmic glue to keep everything from flying apart. It was all theory thrown into the air in the hope that what came down might contain the answer.

The theories failed. The mystery grew to a level of paranoia that we might very well have conclusions about the universe that were so wrong they would be worthless. "After all this time and all this effort," said Charles Alcock, head of astrophysics for Lawrence Livermore National Laboratory, "we still don't now what most of the universe is made of."

Possible answers to this dilemma started off with the theory that the invisible matter equal to more than 20 trillion massive suns is gathered in a very small cluster of galaxies many millions of light-years from the Milky Way. Then came the suggestion that vast clouds of X rays moving with the speed of light permeate the universe, and it is the mass of these X-ray regions that keeps galaxies from coming apart at the seams. How much mass of X rays? Thirty times greater than all the visible matter of the universe.

Those suggestions and theories created their own questions, which in turn only raised more questions. If such density existed on so grand a scale, then the expansion of the universe was definitely limited, and the force of gravity exerted through all spacetime would finally overcome this expansion, stop the outward rush of galaxies, and begin to wind the universe up like a yo-yo coming back along its string.

This would mean a simplistic answer to at least one part of the question: Where did everything come from?

The Big Bang could be accepted for what it was—a theory

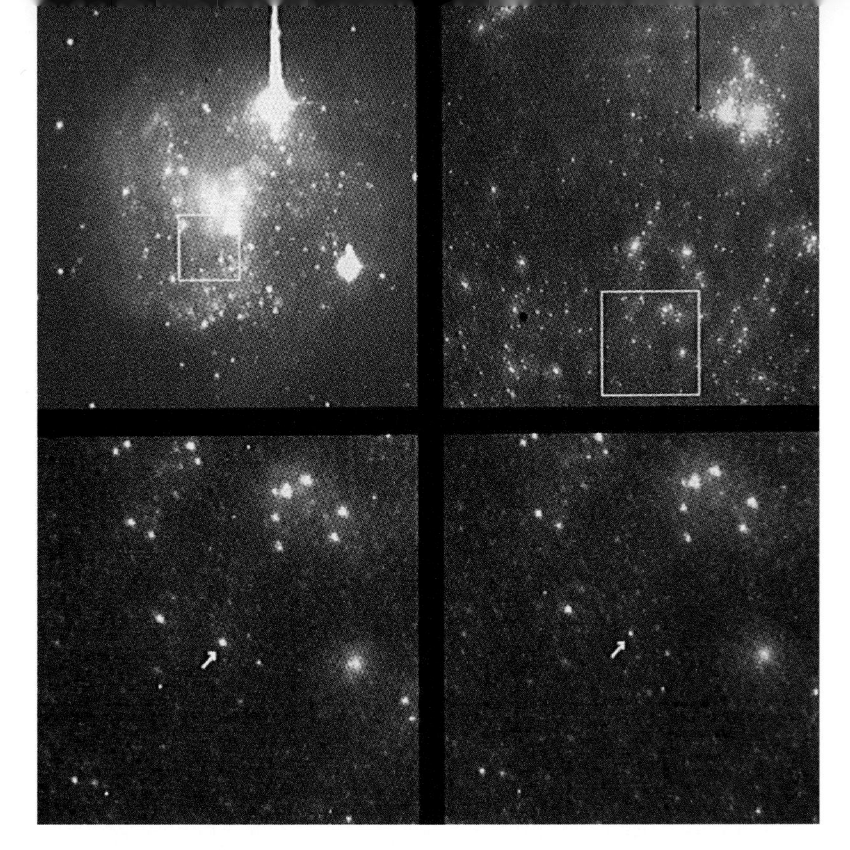

Left: This four-section photograph of Cepheid stars used by Hubble to accurately measure distances in the universe gives us some idea of the many types of stars in the heavens. There are billions of times more stars in our universe than there are grains of sand on our Earth. *(Space Telescope Science Institute and NASA)*

that cannot be proven. But everything that happened afterward would be a matter of a universe expanding to limits imposed by the gravity of dark matter, the collapsing back upon itself until everything that existed was crammed into a tiny point too small to be seen—including all space and time—and finally this contracted mass would explode in the form of a new Big Bang.

Could our universe really be like a huge accordion, alternately expanding and contracting forever? That made as much—or as little—sense as everything else.

Enter Hubble.

No sooner had the huge telescope been repaired by space-walking astronauts in December 1993 than Hubble went on the prowl for the "missing dark matter" that scientists were convinced held the universe in balance. Essentially it would be a uniform expansion with a few bumps and irregularities that didn't affect the overall equilibrium of the cosmos.

The Hubble quest concentrated on finding trillions of stars scattered everywhere through space, so small and so faint they could not possibly be detected by ground telescopes. The red dwarf stars of hundred-billion-year longevity would be found and measured, their mass derived from the Hubble images, and cosmologists would be honored with their accuracy of prediction about dark matter.

By mid-November 1994, the theories went up in smoke. Hubble didn't find anything near the number of red dwarfs postulated

Above: The Beta Pictoris Gas Disk, a near-stellar region roughly the size of our inner solar system. *(Space Telescope Science Institute and NASA)*

by the supporters of dark matter theories. Nor did it find indications that trillions of microsized black holes whizzed about the universe to account for 90 percent of all matter that has defied detection.

"It's fairly embarrassing thing for an astronomer to admit we can't find ninety percent of the mass of the universe," said Bruce Margen of the University of Washington. "It's a constant, ongoing scandal."

The answer would be provided by Hubble's new Wide Field Planetary Camera 2. The random survey of the universe for red dwarfs was a devastating disappointment. Small stars, some with only 10 percent of the mass of our sun, were believed to be the most common stellar bodies in the universe until this moment of Hubble inspection. If they *were* there in great numbers, the majority of all stars, that finding would do very well as a replacement for "dark matter" no one could explain.

Hubble images confirmed that the total mass of small stars accounted for less than seven percent of the mass of the dark-matter halo of the Milky Way.

Scientists aimed Hubble to peer through gaps in the cores of globular star clusters. Hubble could photograph details through the entire cores, and hopes were raised that the images captured by Hubble would show a mass concentration of red dwarf or similar-sized small stars.

Nothing was found with a density less than 20 percent of our sun. Those small stars on which so much hope was placed simply did not exist.

Scientists seemed to be grasping at straws, now postulating that the universe was swarming with massive subatomic particles—called baryons—that had so far proved not only invisible as light but so elusive they had never been found. The new word was that these baryons were made up of still smaller atomic particles. If enough baryons could be found, cosmologists might discover a way out of their now-impossible situation of being unable to find 90 percent of the universe we had studied for so long.

The next offering was that perhaps the universe was filled with electrically neutral, but gravitationally massive, particles called wimps. According to one group of scientists, there are so many wimps racing about and flashing through the universe that all human beings are penetrated by nearly 30,000 wimps every second of their lives. The postulation of wimps would keep alive the dark-matter glue of the universe.

There was only one problem: Wimps are strictly theoretical. They are made up of imagination in a desperate attempt to replace the dark matter no one seems able to find.

So the search goes on while astronomers remain openly stunned at the discovery that they cannot tell us the basic answer as to what makes up the bulk of the universe. Perhaps experts will come up with some new device so that future spacewalkers can place it on board Hubble to answer the question.

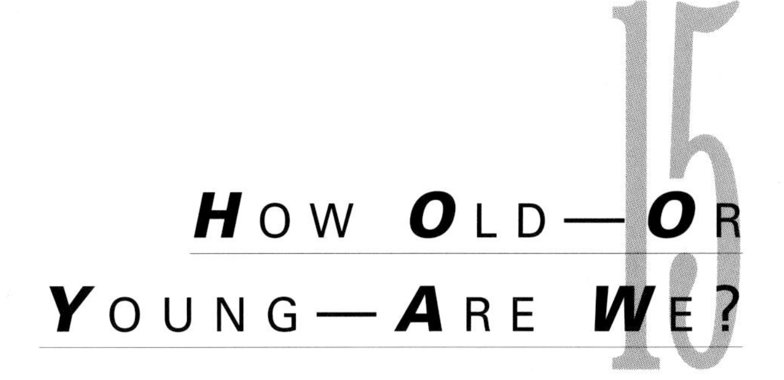

15 How Old—or Young—Are We?

From the first moments when scientists proposed to design and build the Hubble Space Telescope, one mission outweighed all others: to confirm the age of the universe. Hubble, scientists predicted confidently, would travel back in time to image stars and galaxies that first emerged from a universe so hot it was opaque.

The goal, however, was more than simply to determine the age of the universe. From that point, almost everything concerning the makeup and activity of the cosmos could be determined. Going back as close as possible to the Big Bang would provide science with a solid position from which to make additional discoveries.

Before Hubble probed the most ancient universe, scientists already discovered stars from 14 billion to 16 billion years old. Even Hubble had imaged such stars. With good reasoning, then, scientists believed the age of the universe could be measured at 16 billion to 20 billion years.

The results of Hubble's time-travel imaging came with all the success of a lead brick dropping on the toes of the scientific community. According to measurements of Cepheid stars in Galaxy M100, then using that measurement to calculate the age of the uni-

Previous page: Hubble is an orbiting astronomical observatory unparalleled in its capability to take us back billions of years in time. In this photograph we see the Galactic Cluster CL 0939+4713, where island galaxies with hundreds of billions of individual stars appear as if they are only one bright star. We are seeing a part of the universe as it was four billion years ago. *(Space Telescope Science Institute and NASA)*

Right: This Hubble image of a region of the spiral galaxy M100 shows a class of pulsating stars called Cepheid Variables. Though rare, these stars are reliable distance indicators to galaxies. Based on the Hubble observation, the distance to M100 has been measured accurately at 51 million light-years (+/– 6 million light-years), making M100 the farthest object where intergalactic distances have been determined precisely. *(Space Telescope Science Institute and NASA)*

verse, Hubble scientists were astounded to find that their own measurements restricted the universe to an absolute age of no more than 12 billion years. Some members of the Hubble team said even that was too generous, that the best estimates should be given factors of doubt, that the universe could only be measured between 8 billion to 12 billion years.

It didn't make sense. Something is very wrong with how distance and time measurements are made of our universe. Wendy Freedman, of the Carnegie Observatories, and her colleagues stated emphatically that the universe could not be more that eight billion years old.

There was the scorpion in the bottle. If the universe was eight billion years old, then how could there be stars measured accurately at sixteen billion years old? One or the other must be wrong, or both measuring systems must be trashed and replaced with new methods. Or, perhaps, both systems were flawed.

Astronomer George Jacoby of the National Optical Astronomy Observatories in Arizona summed up the issue: "Cosmology is in a time of crisis."

What is even more baffling is that scientists on both sides of the controversy about the age of the universe—the eight-billion-year camp and the twenty-billion-year camp—accept without question the way the ages of stars are measured. Scientists believe those stars judged at sixteen billion years of age *are* of that age, which is, of course, in direct contradiction of the universe age that Hubble appeared to indicate—because Hubble has also confirmed stars aged fourteen to sixteen billion years old!

Once again there loomed the equivalent of dark matter, a "cosmological constant" in the universe that had confounded even Albert Einstein. A cosmological constant, according to scientists, would act as an antigravity force and mess up the measurements made by science of how fast—and how long—it took the universe to reach its present size.

Considering all the measurements and observations they have taken, astronomers hope fervently that the theory of eight billion years is dead wrong. If it isn't, then the favored theories of the Cosmic Egg and the Big Bang will have to be discarded. Something has to give; new answers must be found.

Right now the words of Mark Twain haunt the scientists: "The researches of many commentators have already thrown much darkness on this subject, and it is probable that, if they continue, we shall soon know nothing at all about it."

The measurements of the age of the universe at this moment certainly seem awry, confusing many disciplines of cosmology.

But even in this moment of darkness there's a marvelous sliver, if not quite a ray, of light at the end of the tunnel through which the astronomers now grope their way.

The magic is found in that term "cosmological constant." This is a force that represents antigravity. It has always been judged as utterly impossible; now it is confusion on the age of the universe that may establish that it *is* a natural force.

If this is the final result, then all the confusion and dead ends of measurements of time and distance will have been worth it

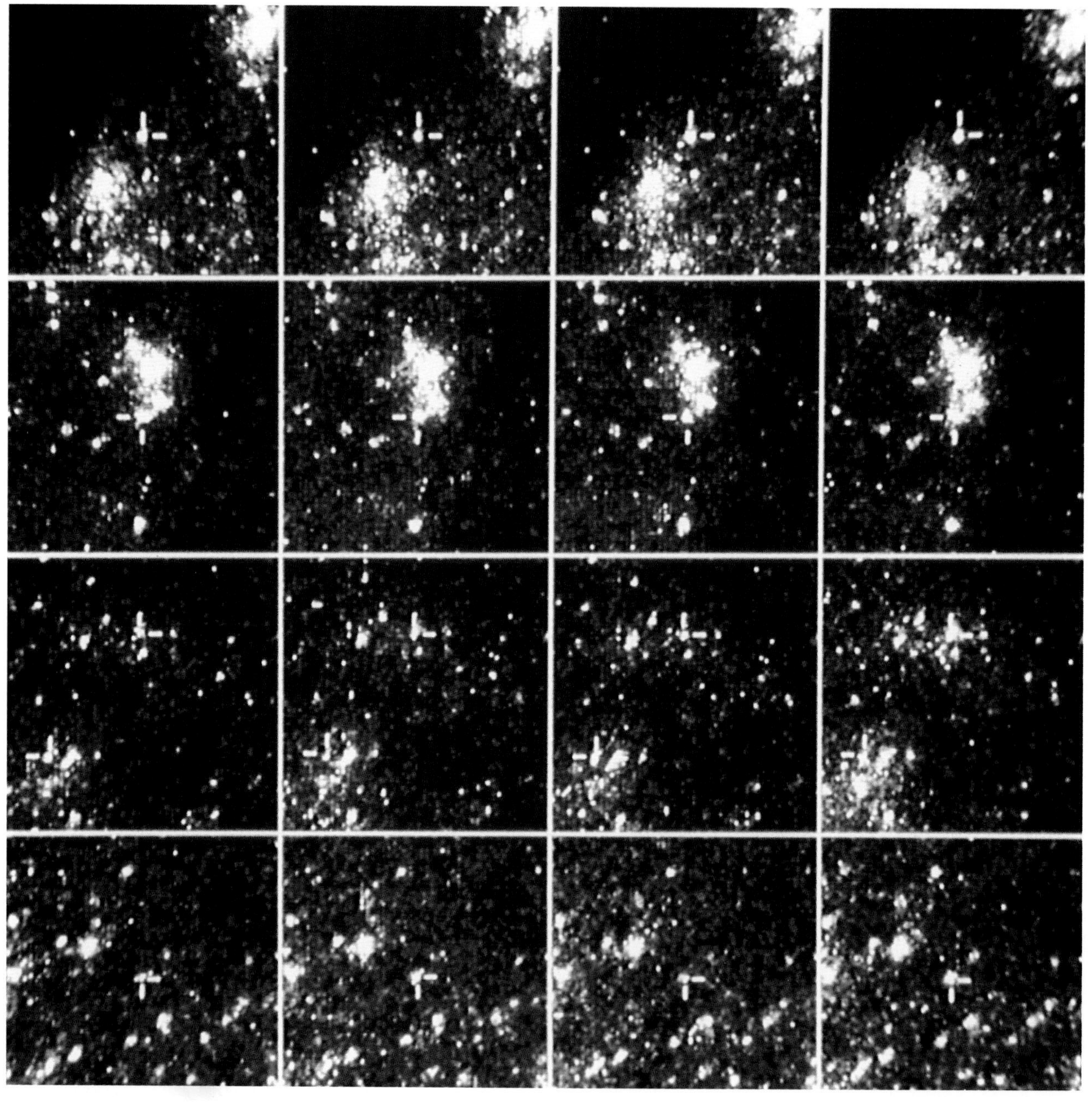

Left: The Hubble Space Telescope has the ability to resolve individual stars in other galaxies and measure accurately the light from very faint stars. This makes the space telescope invaluable for identifying a rare class of pulsating stars called Cepheid Variable stars, as seen here in this collection of photographs taken of galaxy M81. The interval it takes for a Cepheid to complete one pulsation is a direct indication of the star's intrinsic brightness. This value can be used to make a precise measurement of the galaxy's distance. *(Space Telescope Science Institute and NASA)*

Above: This picture by Hubble of an ancient galaxy cluster is one of the faintest and farthest provided by the orbiting space telescope. Astronomers believe with this snapshot of the universe, Hubble is showing us images of stars and galaxies when the universe was an infant. We are seeing a slice of the universe that occurred seven to ten billion years ago, long before Earth and our solar system was born. *(Space Telescope Science Institute and NASA)*

many times over. If an antigravitational force really exists through the universe, every effort will be made to identify it, discover its source, and learn how it works. Above all, scientists will try to learn how to harness a natural energy that can be turned into a quantum leap forward in traveling not only to the other planets of our solar system but beyond and outward to the stars.

The contentious period of accurately defining the ages of stars and of the universe is not likely to endure for long. The Hubble Space Telescope is an instrument of unparalleled capability, and cosmology is a science of discovering new data, subjecting it to intensive tests, and, if necessary, discarding those data in the face of even more accurate definitions.

While scientists' self-doubts and confusion muddy the astronomical waters, no one questions that much of what Hubble (and other instruments of cosmic study) has discovered and confirmed is indeed accurate.

Every time Hubble penetrates hitherto murky areas of nebulae, stars, gaseous clouds, and other areas of near and distant space, another step is made toward defining this astonishing and marvelous universe. Every effort is made, for example, to examine every possible manner in which stars are created and how protostars suddenly burst into life. Theory will do only until those moments when repeated observations change theory to accepted fact—even if theories are revised in the face of the observations. There is no universal law in the cosmos that stars *must* be born in precisely the same way from the raw protostellar materials spanning trillions of miles space.

Let us return to the mainframe of Hubble. By now we—and all the world—are very familiar with Hubble's problems. Few potential disasters rarely leave behind that wreckage of their own faults, yet Hubble did more than that. When the first scheduled servicing and repair mission ended in December 1993, Hubble estab-

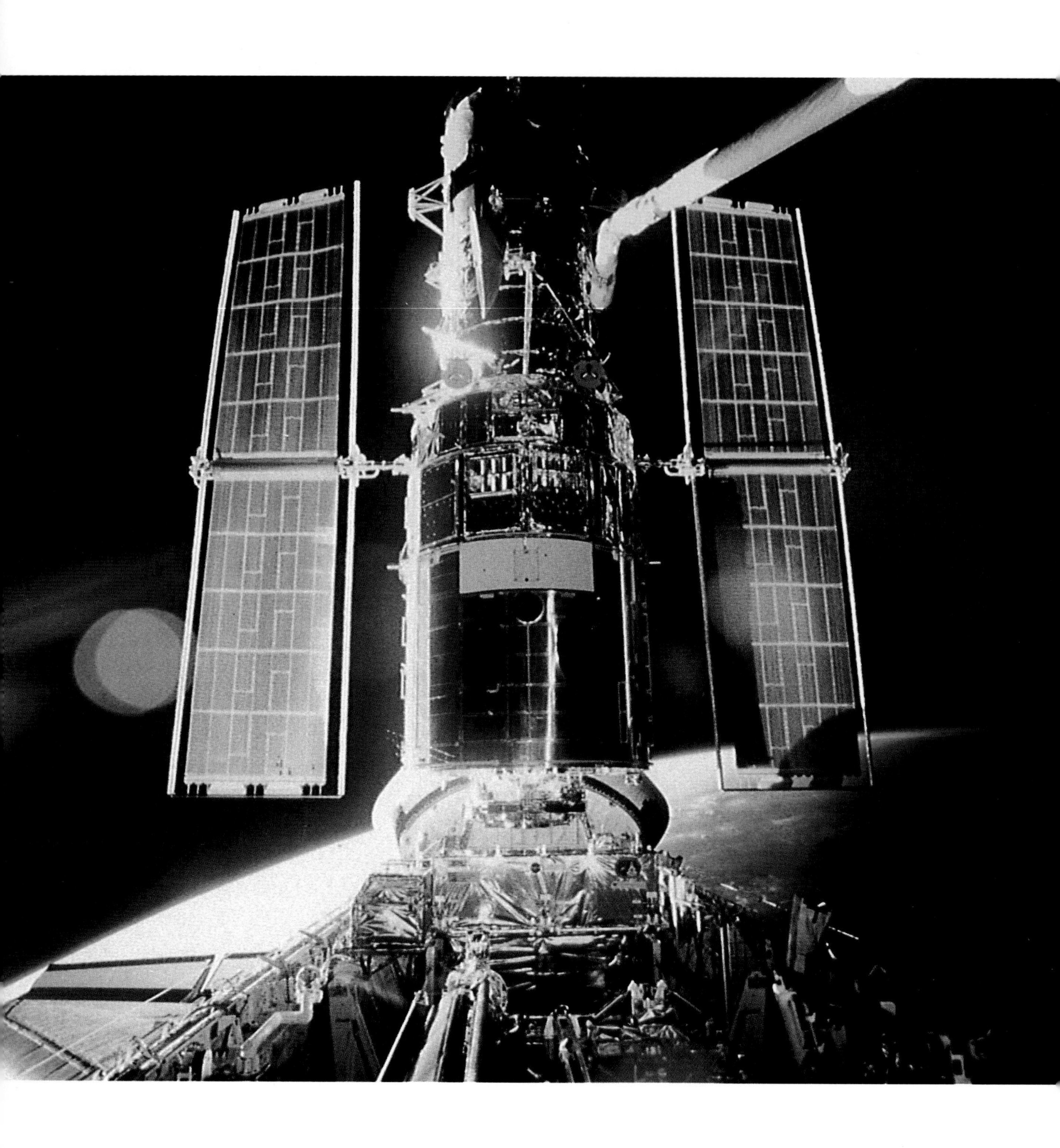

Left: The Hubble Space Telescope is seen here in the grip of the shuttle Endeavour's robot arm as service was completed by space-walking astronauts in December 1993. Hubble has since performed beyond the most optimistic estimates made for the giant telescope prior to its launch. *(NASA)*

lished through its international teams, and extraordinary cooperation with astronomers and astrophysicists through the world, that it was not only better than new, it outperformed the most optimistic expectations.

On December 6, 1994, scientists, technicians, and the news media waited expectantly to hear the results of Hubble's search of the depths and most distant reaches of space in the nationwide news briefing.

By the time the findings were presented, it was obvious the world was ill-prepared for what the Hubble scientists had to say. First, Hubble teams had photographed a universe like nothing ever seen before. Scientists began to identify—this time beyond question—the long-sought population of primeval galaxies that began to form less than one billion years after the Big Bang. It was almost a necessity for the people evaluating the photographic and other evidence that they study not primeval galaxies as they exist today, but as they did when they had surged through an early universe that changed shape so drastically since the light from these formations started on their way to Earth that they had not existed for many billions of years!

Every prediction that Hubble would prove to be the ultimate time machine for such exploration, catching images and fury from a past almost lost in time, came true.

"The researchers," stated scientists and technicians of the Space Telescope Science Institute, and the Association of Universities for Research in Astronomy, allied also with the Goddard Space Flight Center in Maryland and the European Space Agency, "used Hubble as a powerful 'time machine' for probing the dim past. This is the astronomical equivalent of digging through geological strata on Earth. Hubble peers across a large volume of the observable universe and resolves thousands of galaxies from five to twelve billion light-years away. Because their light has taken billions of years to cross the expanding universe, these distant galaxies are 'fossil evidence,' encoded in starlight, of events that happened long ago."

The single most outstanding news of what Hubble discovered is that it encountered stellar formations neither it nor any other astronomical device had ever seen or even suspected of existing.

At first glance the scientists were surprised to discover elliptical galaxies, oval shapes still present in the universe during these last few billion years. But when they looked for the familiar spiral galaxies—wheeling star cities with arms flung outward by rotation from the central galactic mass—they were gone. Fourteen billion

Right: Hubble's solar panels are seen here gathering needed energy from the sun as the space telescope is placed on its orbiting work station by the shuttle Endeavour's crew in December 1993. *(NASA)*

years in the past, much closer to that cataclysmic violence of the Big Bang, it appeared that elliptical galaxies were the first fully evolved stellar cities to achieve the same shape we see today. "These pictures," NASA announced at the December 6 news briefing, "obtained the clearest views yet of distant galaxies that existed when the universe was a fraction of its current age. A series of remarkable pictures, spanning the life history of the cosmos, are providing the first clues to the life histories of galaxies."

In the prevalence of elliptical galaxies and the dearth of spiral forms, the evidence was clear that ellipticals evolved in unusually swift time when raggedy clumps of matter dominated a universe filled with hot gases and outrushing energy.

Yet that presented an immediate problem. Stellar formation, observed from the present to the most distant known past, reveals a universe rich in the great spiral forms—except that now there existed proof that somewhere along the way the first spiral galaxies were devastated by the natural expansion and surging violent forces of the universe.

Scientists remain convinced that spiral galaxies—like our Milky Way and Andromeda—also were molded into existence in the early universe. *But where were they?*

Once again, Hubble photographs snatched definition from contradiction. The most recent Hubble observations clearly identified fully formed elliptical galaxies within a pair of primordial galaxy clusters. The key now was to narrow down the time span when these ellipticals existed. From the Hubble Telescope came images of unprecedented clarity, with sufficient details to confirm long-standing theories.

What surprised the scientists is that the elliptical galaxies in every respect appeared to be remarkably normal in comparison to similar galaxies much younger and closer to Earth. But these pictures captured scenes of the universe when it was but a small fraction of its present age.

The new clues yielded their secrets reluctantly. Turning to our own Milky Way to understand how galaxies evolve provides only limited answers. Although the 400 billion suns of our galaxy are stars in various stages of evolution, the Milky Way itself is a rarity among galaxies—it is still young and vibrant, and thus holds few clues to stellar evolution in very old galaxies. We can see young galaxies with Hubble—or see them when they were young, 8 to 12 billion years ago—but light diffusion, problems with telescope resolution, and a universe filled with vast amounts of gas and dust make it impossible to determine the details astronomers need to

Left: This view of the "Gateway to the distant universe" is arguably the best image yet taken by the space telescope.

The brightest object in this picture is the brilliant galaxy NGC 4881. It is a 13th-magnitude elliptical galaxy in the outskirts of the Coma Cluster, a great cluster of galaxies estimated to lie at a distance of more than 300 million light-years. *(Courtesy William Baum, the HST-WFPC team, and NASA)*

understand how spiral galaxies start, flourish, and end their lives.

This is where clumped, diluted star material, and a ragged and shredding universe, yielded their secrets. The pictures Hubble produced gave strong indications that galaxies formed in the earliest times of the universe accreted from the "lumpy oatmeal gruel" of stellar material rather than from a great mass of matter in space.

This "diluted by lumpy mixture of hydrogen and helium gas—the primordial elements forged in the Big Bang," reported NASA, "also indicate that two vastly different scales of mass prevailed less than 100 million years after the Big Bang, which ultimately affected the formation of galaxies."

NASA scientists faced new and disturbing possibilities of cosmic evolution, advising that "matter was either clumped into vast collections more than a million times the mass of the Milky Way, or into small clumps one million times smaller than the Milky Way."

Then the NASA reports raised eyebrows and left a sinking feeling with those people who'd gathered to be granted the wonders of really learning what the early universe was like in its expansion. As NASA's Dr. Ed Weiler had prophesied, Hubble was going to toss many more questions into the ring that it was going to answer.

Scientists who ran the COBE and other satellite programs confirming that our universe is filled with radiations and forces invisible to the human eye had long before the great search by Hubble virtually thrown out the whole concept of the Big Bang. The COBE report indicated that one instant there was nothing, not even an infinitesimal speck of mathematical singularity, and then, however it may have happened, there was this ultimate release of energy and matter, and in millionths of a second the entire universe sprang, fully formed, into being. Whoever or whatever created this most extraordinary ready-made universe put in place all manner of galaxies and other stellar formations and forces.

But if *that* were true, then how could—or why would—galaxies develop over billions of years from raggedy lumps and clumps into expanding and growing shapes? Both possibilities were so diametrically opposed that both could not be true.

When NASA reported on the latest findings of Hubble, one statement was packed to the brim with *maybes* and *perhapses*, hardly data of a conclusive nature.

"Superclusters of galaxies may have evolved from [vast collections of matter]. . . . Globular clusters—spherical collections of very old, densely packed stars usually found in orbit around galaxies, like the Milky Way—may have evolved from [small clumps of matter]."

NASA groped on. "Could these globular clusters be the meager leftovers of ancient, once-common population of small clumps as predicted by theory? This possibility now seems increasingly more likely. So the question then arises: What formed the vast majority of the galaxies?"

More definitive answers came in the color of light emitted by the most ancient stars. As Hubble punched deeper back through time and cosmic debris, always pursuing the brightest starlight, astronomers detected a change from earlier such observations. They now were well into a cosmic maze, a mystery of almost infinite time and distance. Their first reward came when blue light waxed ever stronger in the most distant galaxies radiating in the visible spectrum. They compared the visible blue light of galaxies close to Earth where blue light was the confirming sign of very young, massive, and extremely brilliant stars forming. *The bridge had been found!* Hubble was showing events that had taken place within only a few short billion years from the tumultuous birth of the first galaxies.

Again star colors yielded fresh clues. The most ancient elliptical galaxies—confirmed at least eight billion and likely twelve or more billion years old—appeared, in NASA's report, as "remarkably normal" when the universe was still a burgeoning, expanding chaos in every direction. Only one conclusion could be drawn from the now-confirmed images of ancient ellipticals: They formed shortly after the sundering violence of the Big Bang.

How, then, did they retain their unmistakable shape and size through time, across more than 12 billion years? The clue lay in the stars' colors. The team led by Space Telescope Science Institute astronomer Mark Dickinson, including lead scientists from private and federal university research programs, concentrated its attention on one particular star cluster that existed when the universe was not yet even one-third its current age.

In that cluster, Dickinson found, to his great surprise, the red galaxies of the early universe resembled quite ordinary contemporary elliptical galaxies. The red color was the long-sought clue, for while blue or blue-white stars definitely are younger, brighter, more powerful stars, the reddish hue from the ellipticals could derive *only* from a population of older stars.

Dickinson and his team exulted in their detailed studies. "This has immediate cosomological implications," Dickinson explained, "since the universe must have been old enough to accommodate them. Cosmologies with high values for the rate of expansion of space (called the *Hubble Constant*, which is needed for calculating the age of the universe) leave little time for these

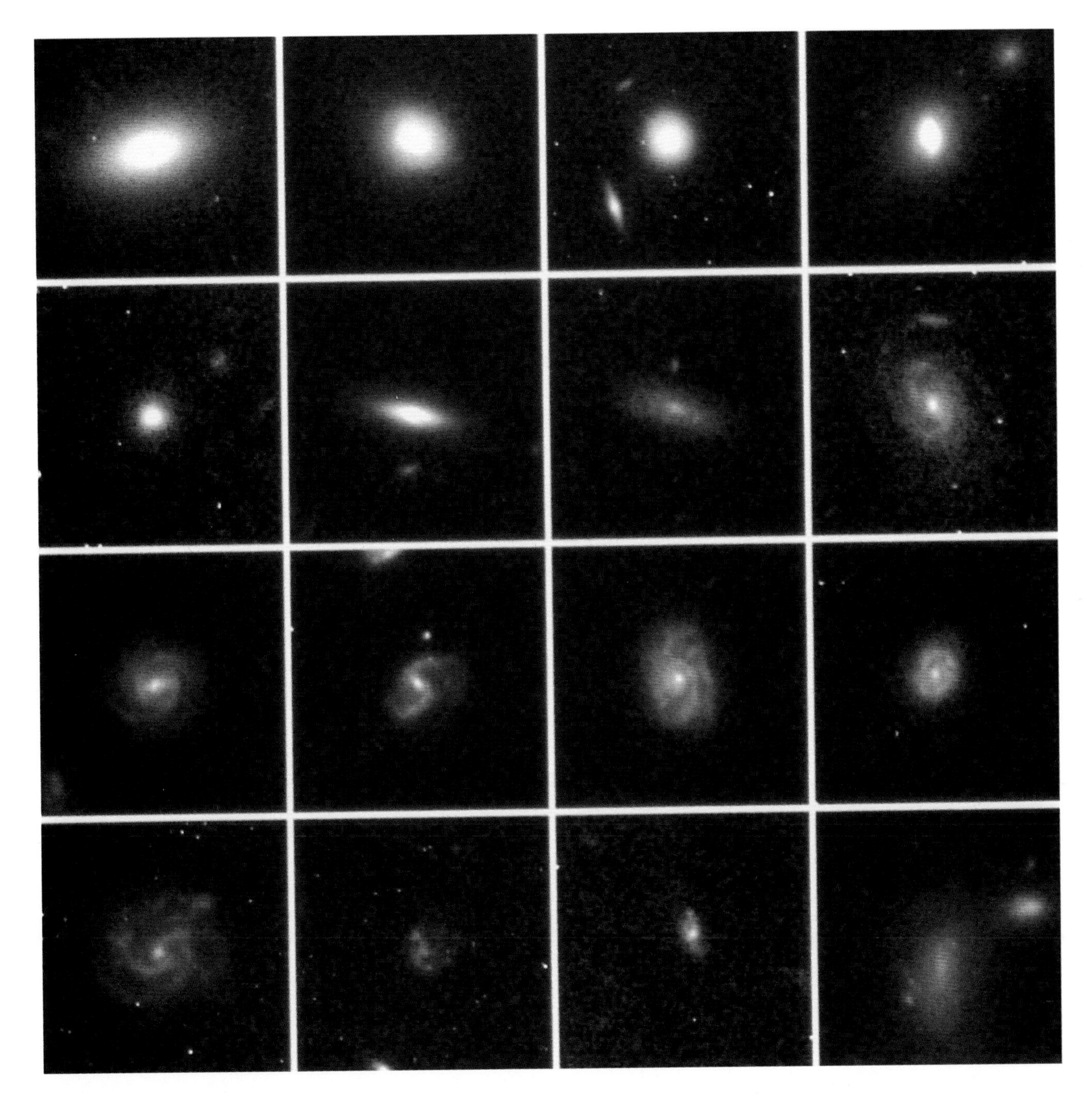

galaxies to form and evolve to the maturity we're seeing in the Hubble Image."

The European Space Agency and Space Telescope Science Institute's Duccio Macchetto led a team of investigators that concentrated on a galaxy with its age confirmed at 12 billion years old. The team pursued the distribution of ancient light that might compare with today's elliptical galaxies. Once again, confirmation was their reward. Macchetto explained that "elliptical galaxies reach their 'mature' shape very quickly, during a robust burst of star formation, and then evolve passively."

Above: This collection of sixteen Hubble photographs was taken of normal spiral galaxies estimated to be five to seven billion light-years away. *(Space Telescope Science Institute and NASA)*

Below: A montage of unusual galaxies discovered during Hubble's deep sky survey. The photos here were taken in visible light (yellow) and near-infrared (red). Galaxies are three to ten billion light-years away. *(Space Telescope Science Institute and NASA)*

Maurio Giavalisco of the Space Telescope Science Institute provided the clincher. "Astronomers suspected that this was the case for at least some ellipticals. Now, Hubble has found direct evidence for it."

The scenario emerging from the new Hubble plumbing of ancient times was that once the elliptical galaxies had formed, they yielded the energetic expansion that characterized their earliest periods. The denser collection of stars in the centers of the ellipticals seemed almost to abandon energetic growth. The great stars shed their outer gases with little or no violence that would create energetic activity through the remainder of those galaxies. It was as if the older, greater stars of the ellipticals settled down to premature old age, burning with a reddish hue and living out long lives, without being affected by whatever else was going on in the universe.

To produce elliptical form in a galaxy required at least a billion years, during which time great masses of gas settled like sinking weights into the center of the galaxy's gravitational field. Another key factor dropped into place. The galaxies we observe as they were only two billion years after the Big Bang followed several definite steps. First, their formation began less than one billion years after the Big Bang. Second, they accreted quickly and produced large families of stars. Third, they are remarkably similar to much younger elliptical galaxies found today in the universe.

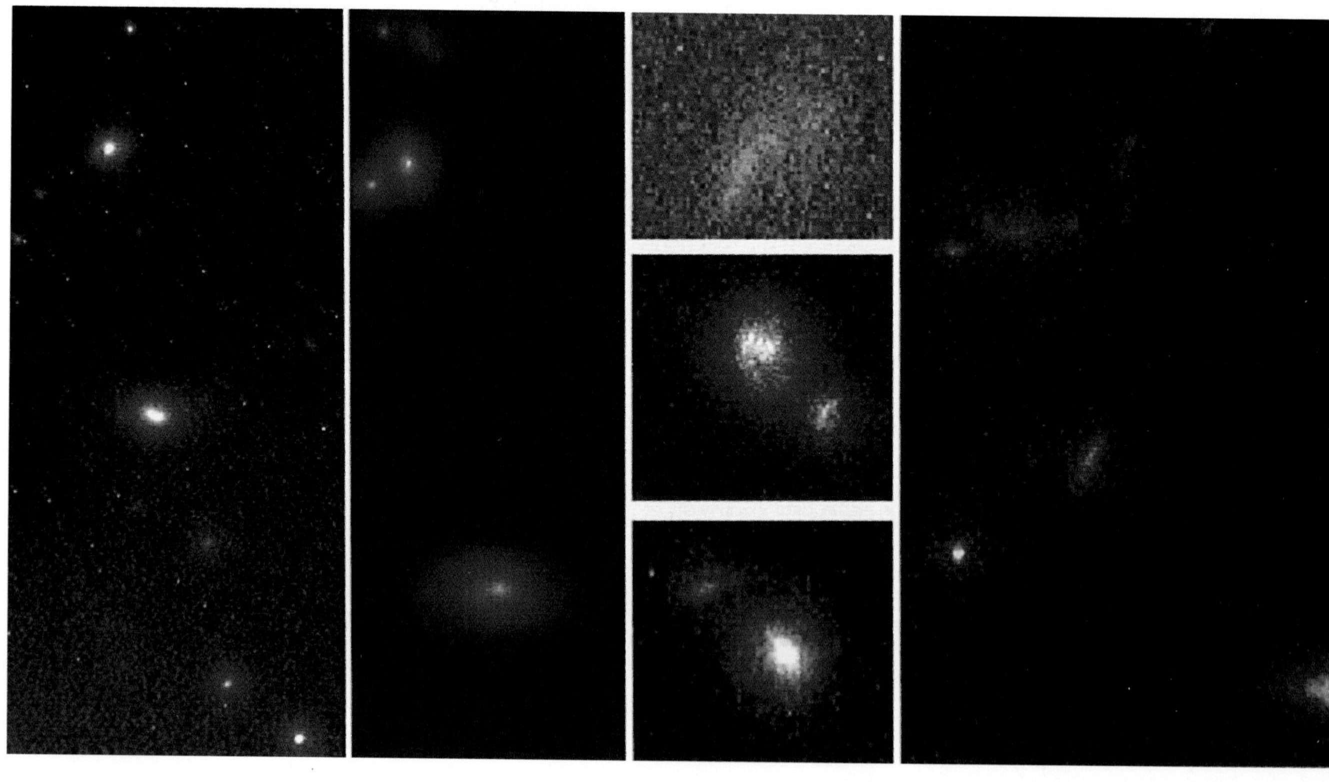

It was as if they had stepped out of the river of time to nearby dry land and stood idly by as the rest of the universe rushed through turbulent and explosive change and growth. As a NASA team concluded, much needs to be learned about the earliest time of the oldest ellipticals. It is too easy to be fooled by short-term observations. But Hubble has given astronomers a new place to stand to comprehend the growth of elliptical galaxies—the sluggards of the universe—which formed during periods NASA now calls "violent relaxation." Everything happened quickly; the stars formed into a dense cluster, and future energetic growth and expansion were simply ignored.

But what happened to the sparkling, explosive, tremendous spiral galaxies?

Three investigating teams led by Alan Dressler, Mark Dickinson, and Duccio Macchetto, with fourteen other scientists, put the Hubble Space Telescope through its most grueling paces to find the answers. With exacting care and exquisite timing, they controlled the attitude and orientation of Hubble for a single image exposure of 18 hours.

The pictures contained within this chapter are as close to magic as anyone will ever reach. The series spans virtually the entire life history of the cosmos. Hubble trapped time in waves of billions of years, snapping its pictures as the eons rolled by. As each image was presented to the scientific teams, they were overwhelmed with the success of their venture. Hubble slowly but surely brought forth the first clues, *ever*, to the history of galaxies from the very rim of ancient time.

One immediate result fascinated and exulted the Hubble teams and sent a shock wave rippling through the other astronomical study teams: The universe could be at least 8 billion years old, but under no circumstances could it be more than 12 billion years of age.

Hubble's images captured early stellar formations, disorganized and never before seen, that went beyond 12 billion to at least 16 billion years back in time. There was the great new rift among scientists. Both groups appear to have carried out meticulous studies and research. But the problem was expressed in the simplest of terms.

Immediate results confirmed that the elliptical galaxies developed rapidly into their present shapes, and these shapes have evolved so slowly they have been almost static.

By this time studying the ellipticals, with the predictions possible, the scientists rushed on to study photographs the likes of

which they had never seen before and could not possibly have forecasted. Before their eyes, in picture after stunning picture, Hubble presented a startling, totally alien universe. It was almost like turning from a photograph of a familiar backyard garden to a primeval forest with never-known terrible creatures, and in those photos was a series that—as the Hubble teams explained in a joint statement—tells a remarkable story of the creation and destruction of spiral galaxies in large clusters.

Look at the photographs of disjointed, twisted, shapeless blobs, intermixing and floating like dimly glowing objects beneath a microscope, and you enter the world of the most ancient of times. Mark Dickinson explains this as a Hubble-revealed "celestial zoo" of faint and compact objects forming the primordial building blocks from which the magnificent spiral galaxies such as our own Milky Way evolved. Irregular bluish fragments are scattered through the photos. These are the twisted and tortured

Left: This sequence of Hubble pictures of remote galaxies offers tantalizing initial clues to the evolution of galaxies in the universe.
Far left column: These are traditional spiral and elliptical-shaped galaxies that make up the two basic classes of island star cities that inhabit the universe we see in our current period in time (14 billion years after the birth of the universe in the Big Bang).
Center left column: These galaxies existed in a rich cluster when the universe was approximately two-thirds its present age (9 billion years after the Big Bang). Elliptical galaxies (top) appear fully evolved because they resemble today's descendants. By contrast, some spirals have a "frothier" appearance, with loosely shaped arms of young star formation.
Center right column: The distinctive spiral structure appears more vague and disrupted in galaxies that existed when the universe was nearly one-third its present age. These objects do not have the symmetry of current-day spirals and contain irregular lumps of starburst activity. However, even this far back toward the beginning of time, the elliptical galaxy (top) is still clearly recognizable. However, the distinction between ellipticals and spirals grows less certain with increasing distance.
Far right column: These extremely remote, primeval objects existed when the universe was nearly one-tenth its present age (2 billion years). The distinction between spiral and elliptical galaxies may well disappear at this early period in time. However, the object in the top frame has the light profile of a mature elliptical galaxy. This implies that ellipticals formed remarkably early in the universe, while spiral galaxies took much longer to form. *(Alan Dressler [Carnegie Institutions of Washington], Mark Dickinson [Space Telescope Science Institute], D. Macchetto [European Space Agency and Space Telescope Science Institute], M. Giavalisco [Space Telescope Science Institute], and NASA)*

shapes of spiral galaxies in their most primitive, yet-unformed shapes, dating back nine billion years. It is Dickinson's strongest opinions, based on this photography and other data, that lead him to believe we are seeing what would be the great spiral galaxies yet to come. Not the same ones we see today, for nature has a peculiarity of temperament with her spiral forms. They develop, expand into the glorious pinwheel shapes with which we are familiar, but remain bent on paths of terrible collision, explosion, mixing, and destruction, tearing themselves to gaseous fragments and shreds, from which new spiral galaxies emerge—the continuous reincarnation of star stuff that gives rise to stars, planets, atmospheres, mastodons, oceans, beetles, and humans.

"We see a bewildering range of galaxy shapes," explained Dickinson. "The Hubble image . . . is where we see a menagerie of strange creatures." Dickinson and his team members are yet frustrated by not having a "direct measurement" of distance. He suspects from all data available that these ragged shapes are also remote cluster members, since they group closely around a distant radio galaxy that also does not resemble anything that can be found anywhere in the presently visible universe.

Among this Dickinson-described "zoo" are "tadpole-like" objects, disturbed and apparently merging systems dubbed "train-wrecks," and multitudes of tiny shards and fragments, faint dwarf galaxies. Or, as Dickinson adds, it is quite possibly a continuing menagerie population of totally unknown objects.

Immediately, Alan Dressler's research program images of Hubble-captured scenarios present several rich clusters. These, Dressler is convinced, chronicle the demise of spiral galaxies inhabiting large clusters. "It seems," he laments, "that almost as soon as Nature builds spiral galaxies in clusters, it begins tearing them apart.

"The cause of this disappearance of spirals from clusters," Dressler continues, "from four billion years ago to the present, is unsettled and vigorously debated. Just the fact that the form of entire galaxies could be altered in so short a time is important in our attempts to find out how galaxies formed in the first place."

Hubble's photographic evidence shows that we are observing large-scale galactic "demolition derbies," Dressler adds. It *could* explain why there were so many more spiral galaxies in rich clusters long ago than there are today. There is left no doubt that, a long time ago, many spiral galaxies were destroyed, or disappeared by flinging themselves apart, or were demolished in collisions that wiped out the spiral shapes. Hubble repeatedly imaged

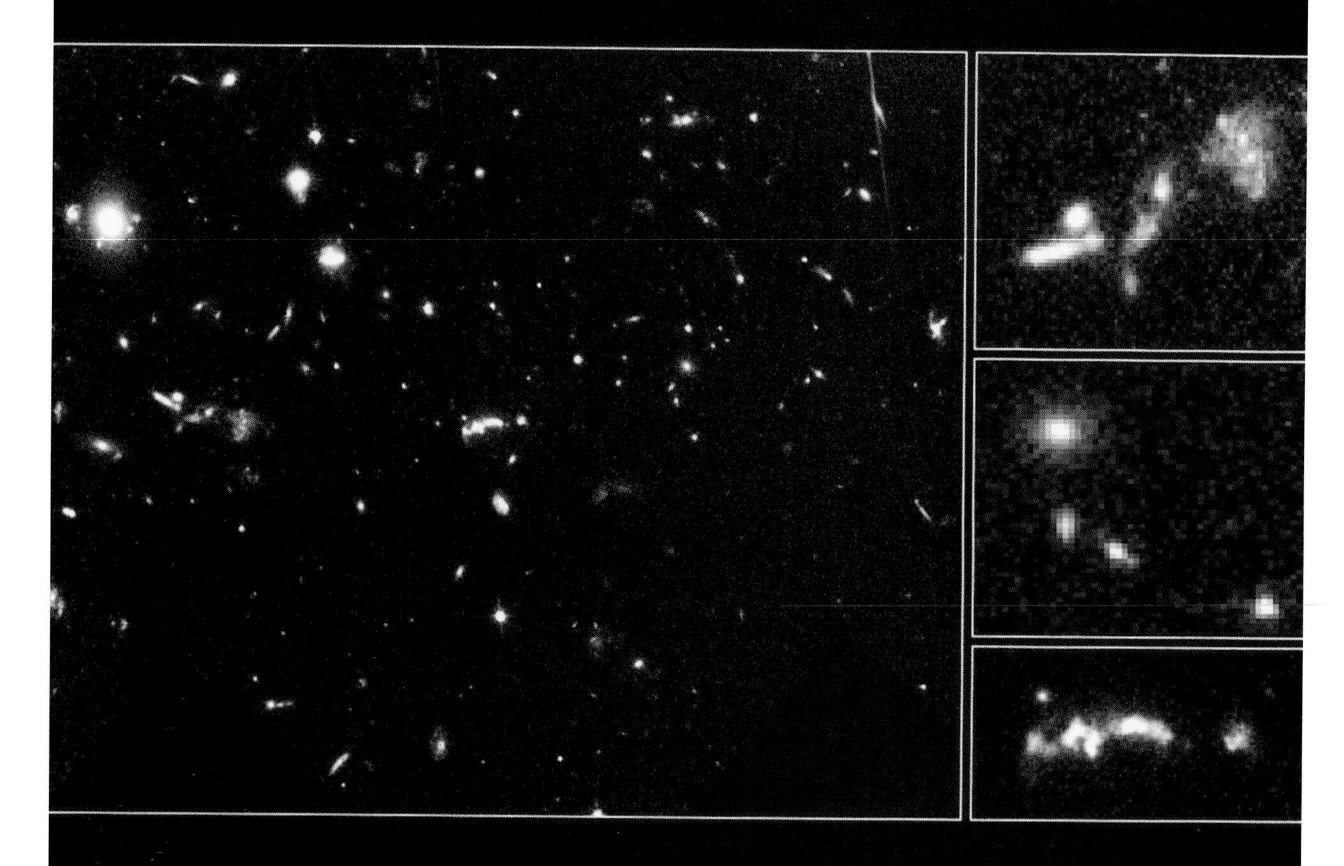

"many unusual objects" within star clusters believed to be torn and ripped fragments of galaxies.

"When we look back in time to these clusters," explains Dressler, "they appear to have been disturbed or disrupted in one way or another. There are so many little shreds of galaxies—it almost looks like galactic debris—flying around in these clusters.

"Perhaps," he stressed carefully, "this is a result of tidal encounters, *but at this point we really don't understand what's happening.*"

The team of Macchetto and Giavalisco concentrated their search on particular quasars, those dazzling cores of very distant active galaxies, to employ the quasar light as beacons to seek out a "shadowing effect" of galaxies between a particular quasar and the Earth. Their reasoning was based on the theory that if a quasar's light is modified (shadowed or dimmed) by a galaxy be-

Left: One of Hubble's deepest images to date of the universe reveals thousands of faint galaxies at the detection limit of present-day telescopes. Peering across a large volume of the observable cosmos, Hubble resolves thousands of galaxies from five to twelve billion light-years away. The light from these remote objects has taken billions of years to cross the expanding universe, making these distant galaxies "fossil evidence" of events that happened long before Earth and our solar system were born.

Very few of the cluster's members are recognizable as normal spiral galaxies (like our Milky Way), although some elongated members might be edge-on disks. *(Space Telescope Science Institute and NASA)*

tween the quasar and Earth, most likely the quasar exists within an extremely primeval stellar cluster.

"All you have to do is look around the quasar using a specially developed optical filter," Macchetto explained, "fine-tuned to observe galaxies at the distance suggested by the change in the quasar's light." Simple enough according to the Hubble team; it worked to their satisfaction. They locked onto Quasar Q0000-263 in the constellation Scorpio. Their long-sought target is the most distant "normal" galaxy ever observed from Earth, 12 billion light-years away.

"The very presence of the cluster shows that . . . large clusters already existed two billion years after the Big Bang," concluded Macchetto. "This is unexpected and counter to many theories of cluster and galaxy formation. Although nothing conclusive can be stated with only one cluster, now that we know how to search for them, we will be able to strongly constrain these theories."

That 18-hour-long exposure of Hubble Telescope developed during 32 orbits revealed thousands of faint galaxies near the very limit of what Hubble can detect with its present power.

Dickinson's team feels an enormous advance in distant viewing by Hubble. The initial results show clearly that mysterious blue cluster galaxies are mostly of the spiral form. Their distorted shapes in the intervening periods suggest they will tell us in future observations why these spiral galaxies vanished "from the present epoch of the universe."

As the Hubble group led by Alan Dressler stated, *"At this point we really don't understand what's happening."*

WHAT LIES BEYOND?

Nobody knows what lies beyond.

They don't even know if there is a beyond.

Within the astronomical community a vast fracturing has built between groups holding tightly to their own cherished beliefs. The teams that insist the universe is only eight billion years old are sharply contradicted by the latest Hubble observations, which, as NASA states, "challenge those estimates for the age of the universe that do not allow enough time for the galaxies to form and evolve in the maturity seen at an early epoch by Space Telescope."

The series of long exposures provided astronomers with detailed pictures when the universe was "approximately one-tenth, one-third, and two-thirds its present age," NASA concluded.

Right: Here, Hubble shows us "Galaxies in the Young Universe."
Left side: This image of a small region of the constellation Sculptor, taken with a ground-based photographic sky survey camera, shows us how difficult it is to see a distant galaxy cluster in the night sky.
Center: Here, Hubble takes a tiny portion of the distant galaxy (inset box), and shows us a snapshot of a slice of the farthest universe as it was 12 billion light-years ago. In this 12-billion-year-old picture of these remote galaxies, this image is a remarkable glimpse of the primeval universe as it looked about two billion years after the Big Bang. The cluster contains 14 galaxies; the other objects are largely foreground galaxies.
Right: This enlargement (small box inset) shows one of the farthest "normal" galaxies (12 billion light-years) yet detected. The galaxy lies 300 million light-years in front of the quasar Q0000-263, which provides a beacon for searching for primordial galaxy clusters. *(Duccio Macchetto [European Space Agency/Space Telescope Science Institute], Mauro Giavalisco [Space Telescope Science Institute], and NASA)*

NASA listed the key findings of the special research effort as follows:

Scientists have begun identifying the long-sought population of primeval galaxies that began to form less than one billion years after the Big Bang.

One of the deepest images of the universe reveals a "cosmic zoo" of bizarre fragmentary objects in a remote cluster that are likely ancestors of our Milky Way.

A series of pictures, showing galaxies at different epochs, offers the most direct evidence to date for dynamic galaxy evolution driven by explosive bursts of star formation, galaxy collisions, and other interactions, which ultimately created and then destroyed many spiral galaxies that inhabited rich clusters.

The researchers use Hubble as a powerful time machine for probing the dim past . . . across a large volume of the observable universe, finding thousands of galaxies from five billion to twelve billion light-years away.

Now that Hubble has clearly shown that it is an exquisite time machine for seeking our cosmic "roots," astronomers are anxious to push back the frontiers of time and space even further.

"Our goal now is to look back further than twelve billion years to see what we are sure will be even more dramatic evidence of galaxies in formation," sums up Alan Dressler.

Was there really a Big Bang when the universe exploded from a cosmic egg that can be described only as a theoretical mathematical singularity? No one knows, because no one knows what a mathematical singularity is. It is a concept based on the idea that something is created from nothing—which is an affront to the logic we know. If there was a Cosmic Egg, where did it come from? No one knows.

If all logic is dismissed, and a Cosmic Egg existed from nothingness, what made it explode? No one knows.

If a Cosmic Egg did explode, why did it explode? No one knows.

How far across is the universe? Some scientists profess to know, but they are busily engaged in fierce arguments with other scientists who have different conclusions.

If the universe is infinite and is constantly expanding, how can it have an edge, or a rim? That is accepting that nothing has an end to it, and that nothing is a special property of nature.

If the universe is finite, where and how does it end? Big problem.

If the most distant galaxies are receding from us at nearly the speed of light, will they reach the speed of light, and then will their mass become infinite? Science states that in our universe nothing can exceed the speed of light, so we have no reason to expect that an island city of trillions of stars will easily violate every law we know and go blithely tearing through some indescribable universe at a speed faster than that of light.

If two separate groups of astronomers are correct in their studies, how can we have stars that are 16 billion years old in a universe that at this moment is deadlocked at a distance of 12 billion light-years? It's a great and continuous bone of contention.

There is acceptance that the universe is dazzling, violent, surging, colorful, explosive, and self-re-creating.

All distances, numbers, quantities, sizes, and times are purely theoretical. The cosmological yardsticks change at an irregular but inescapable pace.

The final conclusion? Keep looking up—for the best seems likely yet to come.

INDEX